Advances in Soil & Forest Research

Dr. S. I. Bhuyan

pencil

ISBN 978-93-5610-348-1
© Dr. S. I. Bhuyan 2022
Published in India 2022 by Pencil

A brand of

One Point Six Technologies Pvt. Ltd.
123, Building J2, Shram Seva Premises,
Wadala Truck Terminal, Wadala (E)
Mumbai 400037, Maharashtra, INDIA
E connect@thepencilapp.com
W www.thepencilapp.com

Author biography

Dr. Shafiqul Islam Bhuyan did his M.sc in Botany from Gauhati University and Ph.D from North Eastern Regional Institute of Science and Technology, Nirjuli, Arunachal Pradesh. He is presently the Assistant Professor & Head of Department of Botany, Pandit Deendayal Upadhyaya Adarsha Mahavidyalaya, Behali, Assam. He published 33 research papers in national international journals, 4 books, 7 book chapters and 3 full papers in proceeding. Guide two PhD students. He also serves as Editor and Reviewer of many Scientific Journals.

CONTENTS

Epigraph

This Book covers different areas of soil characteristics and forest research. Importance is given on enabling the reader to understand the concepts and recent approaches to different areas of soil and forest ecology such as soil health, physical, chemical properties, impact of forest on soils, plant diversity, forest fragmentation, conservation and sustainable development etc.

In the modern world, where the soil and forest resources are increasingly threatened, ecologist and conservationist are playing a major role in providing scientific inputs in conservation efforts. Therefore, there has been a need to compile and collate the research findings in these areas and link bio sciences research with soil and forest ecosystem conservation and sustainable development. This book entitled "Advances in Soil and Forest Research" is an attempt in this direction.

The volume comprises 3 research investigations contributed by eminent scholars of various fields of soil and forest ecology. The book would be excellent references for researchers, policy makers and practitioners.

Preface

Soil properties of forest soils develop under natural conditions by the influence of everlasting vegetation over a long period of time. Physical and chemical properties are almost permanent properties, hardly altered by small factors present in the ecosystems. However, it was found to be changed due to different forest disturbing factors. These properties of forest soils affect every aspect of soil fertility, productivity, root penetration, water availability, water absorption by plants, oxygen and other gases in the soils etc. Therefore, it is clear that soil physical properties mostly affect the plant species richness, their population and forest biomass production in forest. Similarly, soil chemistry determines its ability to supply available plant nutrients, affect microbial processes too. In forest soils more litter with an associated unique micro flora and fauna found as compared to other soils. Forest soils are generally subjected to fewer disturbances than agricultural soils, particularly those that are tilled, so forest soils tend to have better preserved A-horizons than agricultural soils. Soils are very sensitive to the forest disturbances present in a natural forest. Proper assessment and investigations of the forest soil resource can provide significant baseline information on forest health and productivity, especially in the face of continued natural and human disturbance. Different land use pattern will have long-term impact on the atmosphere and climate. Conversion of natural forests into agricultural land (agro-forestry and agricultural land) results in increase of CO2 emission from the soil. Soil characteristics such as soil temperature, soil moisture and soil organic carbon, other soil nutrients are the predominant variables and have an immense impact on soil CO2 flux.

Soil degradation has raised some serious debate, and it is an important issue in the modern era. It refers to the decline in soil's inherent capacity to produce economic goods and perform ecologic functions. Both soil degradation and restoration are mostly affected by different processes which are being regulated by natural and anthropogenic factors. The degree of soil degradation depends on soil's susceptibility to degradative processes, land use, the duration of degradative land use, and the management.

North-eastern hilly region of India is characterized by heavy soil erosion, loss of soil fertility and deforestation causing acute environmental degradation and severe ecological imbalance. In north-east India, the average annual loss of top soil, organic carbon, phosphorus and potassium due to shifting cultivation were to the extent of 40900, 702.9, 0.15 and 7.5 kg per ha, respectively. Different land use systems prevalent in the state and they have the impact on the soil quality and its health. Shifting cultivation in its more traditional and cultural integrated form, is an ecological and economically viable system of agriculture as long as population densities are low and jhum cycles are long enough to maintain soil fertility. However, improvement in physical properties of soil under the humid sub-tropical climate of the northeastern Himalayan region, where erosion is a major land degradative process the locally available natural resources in agro forestry system is a viable option for eco-restoration, maintenance of soil resources and could sustain long-term soil and timber productivity so as to improve food security of the poor tribal farmers of northeast India.

That is why soil conservation is an important issue at present time. Increase the area of natural forest, protection of the plant resources may be the best strategy to protect the soil health and maintain the productivity.

The book is prepared with all the best efforts to make it rich and upto date. We would appreciate receiving valuable suggestions and constructive criticisms relevant to this edition.

3. A Comparative study of Soil Characteristics Found in Different Land use systems of Ziro Valley, Arunachal Pradesh

A COMPARATIVE STUDY OF SOIL CHARACTERISTICS FOUND IN DIFFERENT LAND USE SYSTEMS OF ZIRO VALLEY, ARUNACHAL PRADESH

Hage Yama, I. Laskar and S.I. Bhuyan

Department of Botany,

Pandit Deendayal Upadhyaya Adarsha Mahavidyalaya-Behali, Assam, India

Email: safibhuyan@gmail.com

Abstract

Soil characteristics control relationship between soil constituents and plant productivity. Soil and vegetation has complex interrelationships because they develop together over again restoration of nutrients to the soil properties. Land use pattern prevalent in Lower Subansiri district viz., Settle cultivation, Evergreen/Semi evergreen forest (Dense), Degraded/Scrub land, Forest Plantations, Current and Abandoned shifting cultivation, home gardens and Grassland et.. They affect the soil quality in different ways. Significant variations in soil parameters in different land use systems were found in the present investigation. Finding of this investigation may be useful in developing proper management strategies of soil health and quality for better production in the district or other area having the similar agro-climatic conditions.

Keywords: Conservation, Mineralization, Nutrients dynamics, Productivity, Soil health,.

INTRODUCTION

Soil can be defined as the solid material on the Earth's surface that results from the interaction of weathering and biological activity on the parent material or underlying hard rock. Soil itself is not essential to a growing plant. Soil serves as a storage reservoir for nutrients and water needed for plant growth. These soil properties are essential to crop production. Soil is a mixture of minerals, organic matter, gases, liquids, and countless organisms that together support life on Earth. Soil is a natural body called the pedosphere which has four important functions: it is a medium for plant growth; it is a means of water storage, supply and purification; it is a modifier of Earth's atmosphere; it is a habitat for organisms; all of which, in turn, modify the soil. Soil is called the "skin of the Earth" and interfaces with its lithosphere, hydrosphere, atmosphere, and biosphere. The term pedolith, used commonly to refer to the soil, literally translates 'level stone'. Soil consists of a solid phase of minerals and organic matter, as well as a porous phase that holds gases and water. Accordingly, soils are often treated as a three-state system of solids, liquids, and gases.

Soil is a product of the influence of the climate, relief (elevation, orientation, and slope of terrain), organisms, and its parent materials (original minerals) interacting over time. Soil continually undergoes development by way of numerous physical, chemical and biological processes, which include weathering with associated erosion.

A number of physio-chemical changes occur on submergence of a soil. Of these the more important are the increase in pH, increase in specific conductance and decrease in oxidation-reduction potential. Soil properties of the solid phase result in both marked variability of water contents and a varying degree of resistance to the elimination of moisture. Soil with an extreme PH is strongly influenced by salts, resulting in very acid sulphated soils to highly alkaline carbonate soils. Soil acidity is enhanced by different factors such as mineral leaching, decomposition of acidic plants, industrial wastes, acid rains and certain forms of microbiology activity. However, many ions, sodium, potassium, magnesium and calcium are responsible to be alkality of the soil. Physiochemical characteristics of soil vary in space and time which is greatly affected by variation in topography, climate, physical weathering process, vegetation cover, microbial activities and several other biotic and abiotic variables. Forest cover plays a significant role in soil formation (Chapman and Reiss 1992). For example, litter fall influences physiological characteristics of soil such as water holding capacity, soil texture, ph of soil and nutrient availability (Johnson, 1986).

The variation of soil properties in space and time Implies that soils has varying capacity to retain and supply nutrients to rice crop. This makes it difficult to correctly manage field input applications.

Soil characteristics control relationship between soil constituents and plant productivity (Khattak, 1996). Soil and vegetation has complex interrelationships because they develop together over again restoration of nutrients to the soil properties (Singh et al.1986). Concentration of micro elements in the soil is a good indicator of their availability to plants. Their concentration gives good information towards the knowledge of bio-chemical cycle in the soil-plant ecosystem (Pundit and Thampan, 1988). However, natural vegetation have an important effect quality than most other plant ecosystem types, through the input of litter with high lignin content, high total net primary production, high water and nutrient demand(Brinkley and Guardian, 1998). Even plant species diversity also affects soil properties as well as soil fertility (Augusto et al.2002). The properties of soil are the important factor for the growth of the plants. The adequate theoretical and practical knowledge of various soils is therefore necessary to study.

The growth of plant community cannot be understood without the knowledge of soil property. Plant growth and yield depend on the physical, chemical, and biological properties of soil. Soil physical properties affect the germination and emergence of the young seedling, root penetration, water movement, soil air composition and availability of plant nutrients.

Crop production potential is greatly influenced by the phys-ical and chemical properties of soils. These same properties also influence related activities such as tillage, erosion, drainage, and irrigation. Important agronomic soil properties include the soil water-holding capacity, infiltration rate, aggregation, temperature, organic matter content, and nutri-ent availability. Soil is a complex system where different factors such as physical, chemical, and biological are held in dynamic equilibrium (Kizikaya and Bayrakli, 2005). It consists of water, minerals, air and soil organic matter. It is often classified in terms of their physical properties like texture, structure porosity and fraction of pore space, affected by composition and proportion of the soil components. On the other hand, these properties air and water movement in the soil and thus the soil's ability to function. The structure of the soil are bound together to form aggregates. Soil texture and texture classes are sand, silt; clay and loam denote the proportion of various particle sizes that it contains a mixture of all particle sizes.

Land use pattern prevalent in Lower Subansiri district viz., Settle cultivation, Evergreen/Semi evergreen forest (Dense), Degraded/Scrub land, Forest Plantations, Current and Abandoned shifting cultivation, home gardens and Grassland. Apatanis are involved with settled rice farming practices along with the well integrated pisciculture. Indeed, the Apatani settled farming system is better studied and documented. Evergreen/semi-evergreen natural forest patches and Pine plantations not only are a source of timber but also are a source of organic residues for soil fertility management of agricultural systems. The river Kalle, which coming down from the hills serves as the primary source of water for the wet rice cultivation system; the water coming down also brings in nutrients which enriches the soil down below. Since the river is a perennial water source, the people exploit and utilize the available water for proper growth of crop as well as the survival of the fishes. Also, the river brings about soil rich in natural fertilizers that are the only source of manure for the cultivation purpose. The pine forest is one of the human-managed systems organized as privately owned forest, clan forest, or community forest. This forest is found mostly at the foothills of the valley adjacent to the settled farmland of the community locally called as 'Sartii'. This plantation meets various requirements of the indigenous such as timber, planks, poles, fuel etc. The common pine species found are Pinus wallichinana and P. longifolia. Bamboo forests is another important component of the Apatani cultural landscape; being an important resource for the community, this culturally valued resource is raised closer to human habitations and is locally known as 'Biije', raised as a monoculture of Phyllostachys bamboosoides; bamboo forests are common and is of multi-purpose value for the locals. Sub-tropical mixed natural Pine forests occur between altitudes of 1000 m to 1800 m. Pinus kesia, P. roxburghii, P. wallichiana and P.merkusii, are found along with other tree species, such as, Alnus sps.,and Elaeocarpus sps. The secondary natural forest patches closer to the villages are under high anthropogenic pressure.

Keeping above in the consideration, present study was undertaken to understand the physical properties of soil under different agro-ecosystems of the district.

The present work is taken up with the following objectives:

The Objectives are

To study the prominent land use system in the Ziro valley under Lower Subansiri district,   Arunachal Pradesh.

To study the details physical properties of the different land use systems.

Finding of this investigation may be useful in developing proper management strategies of soil health and quality for better production in the district or other area having the similar agro-climatic conditions.

REVIEW OF LITERATURE

Soil quality is how well soil does what we want it to do. More specifically, soil quality is the capacity of a specific kind of soil to function, within natural or managed ecosystem boundaries, to sustain plant and animal productivity, maintain or enhance water and air quality, and support human health and habitation (Karlen et al., 1997). Soil quality also depends on the soil formation. The idea is that if all five of the soil-forming factors are the same, then the soil will be the same. The technical term used for soil formation is pedogenesis. (Gershuny, Grace. 1993). Soil quality is "the capacity of soil to function within ecosystem boundaries to sustain biological productivity, maintain environmental quality, and promote plant and animal health (Doran and Parkin, 1994). The capacity of a soil to function within its ecosystem boundaries and interact positively with the environment external to that ecosystem. The state of existence of soil relative to a standard, or in terms of a degree of excellence. The continued capacity of soil to function as a vital living system, within ecosystem and land-use boundaries, to sustain biological productivity, maintain the quality of air and water environments, and promote plant, animal, and human health. (Doran et al. 1996). Variation in soil physico-chemical properties in different fields might be due to dynamic interactions among environmental factors such as climate, parent material, topography and land cover/land use (Dengiz et al. 2006).At one time it was thought that soils were static.

Soil development is caused by climate and living matter acting upon parent material (weathered mineral or organic matter from which the soil develops), as conditioned by topography, over time (Brady and Weil, 2002). Soil structure is the arrangement and binding together of soil particles into larger clusters, called aggregates or 'peds.' Aggregation is important for increasing stability against erosion, for maintaining porosity and soil water movement, and for improving fertility and carbon sequestration in the soil (Nichols et al., 2004).

Agriculture land use

The agricultural sector continues to play a crucial role for development, especially in low-income countries where the sector is large both in terms of aggregate income and total labor force. Agriculture is the main source of livelihood for 86 percent of these rural households. Some 75 percent of poor people still live in rural areas and derive the major part of their income from the agricultural sector and related activities. Agriculture provides food, income and jobs and hence can be an engine of growth in agriculture-based developing countries. In Schultz's view, agriculture is important for economic growth in the sense that it guarantees subsistence for society without which growth is not possible in the first place. This early view on the role of agriculture in economics also matched the empirical observation made by Kuznets (1966) that the importance of the agricultural sector declines with economic development. In-dian agriculture is now poised for technical transformation for ensuring food security, export earnings and decentralized development to reduce rural poverty owing to the severe population pressure on the natural resources base of land, water, bio-diversity and other resources to meet its growing food and development demands (Wani M.H. et.al. , 2009).

But in recent year's uses of agricultural land or area of agricultural productivity is being declined, because the action and interaction of various factors such as population pressure, socioeconomic forces, live stock pressure and various types of institutional development that regulate the land use in formally and informally (Shasi Chawla, 2012).

In a dynamic world, certain modification can oc-cur in the existing pattern of land utilization (Lekhi R.K. & Jogin-dre Singh, 2011). In-dian agriculture is now poised for technical transformation for ensuring food security, export earnings and decentralized development to reduce rural poverty owing to the severe population pressure on the natural resources base of land, water, bio-diversity and other resources to meet its growing food and development demands (Wani et.al. , 2009). But in recent year's uses of agricultural land or area of agricultural productivity is being declined, because the action and interaction of various factors such as population pressure, socioeconomic forces, live stock pressure and various types of institutional development that regulate the land use in formally and informally (Shasi Chawla, 2012).

Land use is a product of interactions between cultural back-grounds, state and physical needs of the society with the natural potential of land (Karwariya and Goyal, 2011). Land use of any region expresses the interaction of the operation of the whole range of environmental factors modified by the socio-economic and historical elements (Narkhed & Gatade, 2010). Desirable land use pattern could be achieved through sectoral plan linkages and there is a need to apply modern sci-ence and technology to enhance productivity on a sustainable basis (Wani et.al. , 2009). In general, the land use pattern indicates the way in which the land area used under various circumstances. The pattern of land use of a country at any particular time is determined by the combination of economic and institutional framework. Hence, the land use pattern and the trends during years will help to suggest the scope for planned shift in the pattern.

Different types of agricultural practices are prevalent in north eastern India and the soils of these fields are managed in different ways since it is not only related to economic condition but also to the social and cultural faith of indigenous people (Bhuyan, 2012). Different land use patterns and continuous cultivation are responsible for the variation in the level of organic matter input and subsequent loss of soil organic carbon (SOC) from the agricultural lands leading to alteration of microbial biomass (Srivastava and Singh, 1989).

On the basis of organic matter content, soils are characterized as mineral or organic. Soil is a living, dynamic ecosystem. Healthy soil is teeming with microscopic and larger organisms that perform many vital functions including converting dead and decaying matter as well as minerals to plant nutrients. Nutrient exchanges between organic matter, water and soil are essential to soil fertility and need to be maintained for sustainable production purposes. Where the soil is exploited for crop production without restoring the organic matter and nutrient contents and maintaining a good structure, the nutrient cycles are broken, soil fertility declines and the balance in the agro-ecosystem is destroyed.

Agriculture is the highest economic sector in the state Arunachal Pradesh, where more than 80% of the people are engaged in this practice. Shifting cultivation (Jhum), terrace farming and different settled cultivation are the main agricultural practices in this state. 55% of total agriculture area is under Jhum, which is practiced by the indigenous hill people since the time immemorial. Traditional faith and culture of the indigenous people are greatly concerns with the agricultural practices. However, extensive practice of Jhum with short Jhum cycle led to soil erosion and loss of soil fertility (Bhuyan, 2012). Thus, land use pattern as well as agricultural management has a significant impact on soil characteristics such which consecutively reflect the soil fertility status of a given area (Singh et al., 2009).

Agricultural landuse is the result of inter-action between man and environment. Besides physical factors such as relief, climate and soils, agricultural landuse are also affected by socio–economic and technological factors. The term "Agricultural Landuse" denotes the extent of the gross cropped area during the agricultural year under various crops. It is the result of the decision made by the farmers regarding the choice of crops and methods for production. Thus, this decision making is based on not only physical constraints and limitations but also on farmer's perception of the total environment. The physical as well as cultural environment affects on crop growth and production.(Vaidya B.C.).It is also related to the changes undergone in the spatial distribution of agricultural use of land, which is result of the direct application of efforts to the available land resource.

Agriculture plays an active role in economic growth through production and consumption linkages. Other precursors, such as Schultz (1964), also point out the importance of food supply by the agricultural sector. In Schultz's view, agriculture is important for economic growth in the sense that it guarantees subsistence for society, without which growth is not possible. This early view on the role of agriculture in economics matched Kuznets' (1966) empirical observation that the importance of the agricultural sector declines with economic development. Integrated paddy cum fish cultivation was done where; the rice field can be utilized for fish culture in the following two ways. Fishes can be reared from the month of April to September when the paddy crops grow in the field. At present it is being practised at Ziro .The fish culture can also be taken up from the month of November to February after harvesting of paddy crops is completed and transplantation for the next season begins. The culture of fishes in paddy fields, also remains flooded even after the paddy is harvested. Giri (1966) studied the changes in land use pattern in India during the period of 13 years, that is, from 1950-51 to 1963-64 under the 1st, 2nd and three years of 3rd five year plan and found that in the 3rd plan the net sown area, current fallow and area under non-agricultural uses increased by 14.7 per cent, 3.7 per cent and 29.5 per cent respectively. Whereas, forest area, barren & uncultivated land, old fallow and culturable waste showed decrease of 0.7 per cent, 1.4 per cent, 30.7 per cent and 29.4 per cent 38 respectively. The cropped area also had changed its distribution widely and considerable acreage had been converted from forest or bush area to crop area.

Singh and Vasisht (1997) analyzed the dynamics of land use pattern in Bihar, Punjab and at All India level during the period 1950-51 to 1990-91 and found that the barren and uncultivated land had increased in Bihar from 1960-61. This increasing trend in land rendered unable for cultivation, had served as a deterrent in the development of agriculture. Agyepong and Sosthenes (2003) examined the spatial patterns of land use and cover as a basis for the analysis of the socio-economic causes of the change in the environment and environmental consequences of land use and cover change in Ghana. The forgoing review of literature reveals that only limited studies have been conducted on physical properties in different land use pattern of the state in general and Ziro valley in particular. Therefore, a proper investigation is important on soil characteristics with particular emphasis to agro-climate conditions in Ziro valley under lower Subansiri district, Arunachal Pradesh.

MATERIALS AND METHODS

Study area

Present study was conducted in ziro under Lower Subansiri district of Arunachal Pradesh which lies between Latitude of 27°33'59"N and 93°49'53"E. Ziro is the headquarter of the Lower Subansiri district and is situated at a height of approximately 1524 meters above mean sea level which lies at 26o 50 N- 98o 21 N Latitude and 92o 40 E and 94o 21 E longitude. There are 6 administrative circles in this district, namely, Ziro (Sadar), Yachuli, Pistana, Raga, Kamporijo and Dollungmukh. The district is also divided into 3 blocks: Ziro-I, Ziro-II and Tamen-Raga. Ziro covers an area of 10,135 km2. It is located at an elevation of 1688 metres (5538 feet) to 2438 meters (8000 feet). The Lower Subansiri district has largest area under permanent agriculture cultivation among all the districts of the State. The Ziro valley is also known as Apatani valley under the Sub-Himalayas climatic zone. Ziro is around 115 km from the state capital Itanagar and is well connected with North Lakhimpur in Assam near the Arunachal-Assam border which is around 100 km from Ziro.

Climate and rainfall

The climatic condition of the district varies from place to place as well as season to season. The climate is largely influenced by the nature of terrain depending upon altitude and location of place. It may broadly be divided into four seasons in a year: The cold weather season is from December to February, March to May is the pre-monsoon season of thunderstorms, the south–west monsoon from June to about the middle of October, the second half of October to November, which constitutes the post-monsoon or the retreating monsoon period and is a period of transition. In the foothills or low high belt area of the district, the climatic condition is moderate in comparison to high belt areas, where during winter is very cold and chill, and in summer is pleasant. December and January are generally the coldest month, and July and August are warmest months. Annual rainfall in the south is heavier than that in the northern areas of the district. During the monsoon period more than 70 percent of the rain over the southern half occurs while in the northern portions it is about 60 percent. Variability of rain fall for the monsoon and the year, as a whole, are relatively small. Average annual rainfall of the district headquarters, Ziro recorded as 934.88 cm during 2000.

Soil condition

The soil developed in each physiographic unit has their distinct morphological and other related properties. It indicates a good soil land form relationship in this region. The soil in the valley area are Loamy sand , Loam,Clay loam with medium texture and is moderately acidic and rich in organic matter content .

Figure 1. Map of the study area

Demography

According to the 2011 census Lower Subansiri district has a population of 82,839. This gives it a ranking of 623rd in India (out of a total of 640). The district has a population density of 24 inhabitants per square kilometer (62/sq mi). Its population growth rate over the decade 2001–2011 was 48.65%. Lower Subansiri has a sex ratio of 975 females for every 1000 males, and a literacy rate of 76.33%. This district is inhabited by Nyishis and Apatanis. Languages used in the district include Apatani, Nyishi language.In ziro they use the apatani dialect in majority and very few minority use nyishi dialect.The main occupation of the population in ziro is agriculture. As of 2001 India census, Ziro had a population of 12,289. Males constitute 52% of the population and females 48%. In Ziro, 17% of the population is under 6 years of age. The main occupation of the population in ziro is agriculture.

Soil sampling

The soil samples were collected from four different selected land use pattern of Ziro under lower subansiri district such as paddy field, home garden, bamboo garden and forest. First, the detritus (grass and litter) at each sampling were cleared from ground surface and the soil was dug with a spade and a dao. The soil samples were collected from three locations of each LUP randomly from the two depths i.e. the upper 0-15 and the lower 15-30cm.

Laboratory analysis

The samples were sealed in poly bags and were brought to the laboratory. In the laboratory the samples were air dried at room temperature, disaggregated by using a wooden mortar and sieved through a 2 mm mesh for further use and analysis. The following parameters were then studied under four different land use pattern and they are as follows.

Water holding capacity

It was investigated with the method outlined by Allen et al. (1974). It can be calculated by the applying formula –

Water holding capacity=Water retained by soilWeight of soil x 100

Moisture Content

Soil moisture content is determined by Anderson and Ingram (1993). Moisture content in soil can be determined by using the formula which needs weight of the soil before heating (x) and weight of the soil after heating (x1).

Percentage of moisture content = X – X11 x 100

X1

Soil texture

Soil texture was determined by Boyoucous hydrometric method given by Allen et al. (1974).

The percentage of sand, silt and clay were determined by using the USDA texture classification chart which is as follows –

C1=R1-R3

C2=R2-R3

C3=C1-C2

Where, C1 = mass of silt and clay in suspension at 40secs,

C2 = mass of clay in suspension at2hr,

C3 = silt portion

R1 = density of soil suspension at 40s,

R2 = density of soil suspension at 3hrs and

R3 = density of soil suspension of blank solution.

Clay % = C2X2

Silt % = C3X2

Sand % = 100 – (Clay % + Silt %)

Figure 2. A soil texture triangle to determine soil texture class from the percentage of sand, silt and clay in the soil.

The data was analyzed and significant error was calculated using Zar, (1974).

Particle density

Soil particle density was determined with the method outlined by Allen et al. (1974).

Particle density can be calculated by using formula as

Weight of water displace by soil = (W2+10)-W3

Particle density =   10

(W2+10)-W3

Where, W1 = Weight of empty pycnometer (R.D.bottle)

W2 = Weight of empty R.D. bottle + water

W3 = Weight of R.D. bottle + soil water

Bulk density

Bulk density was determined using the core method as described by Anderson and Ingram (1993). Bulk density was calculated as follows –

Weight of soil = W2 – W1

Volume of soil = V

Bulk density = (W2 – W1)

$$V$$

Where, W1g = Weight of empty bottle

W2g = Weight of bottle + soil

V ml = Volume of soil or volume of water needed to fill the bottle

Porosity

Porosity refers to the volume of soil void that can be filled by water and air. It is inversely related to the bulk density. It was determined by indirectly from the value of bulk density and particle density. Soil porosity was determined and outlined by Allen et al. (1974).

Porosity is pore volume expressed as a fraction of total soil volume. Porosity can be calculated if bulk and particle densities are known.

Porosity = 1 - (bulk density / particle density) x 10

Statistical analysis

The data on soil were analyzed using ANOVA to study the different land used pattern, sampling period and soil depth on different properties of soils and their changes. Correlation analysis was completed following Zar (1974) to study the relationship between soil characteristics and microbial biomass.

RESULTS AND DISCUSSION

The physical characteristics of different soil samples from selected agricultural land in ziro under lower Subansiri district, Arunachal Pradesh have been studied and the data were summarized in the tables and figures mentioned below.

Physical properties

Soil Moisture content

The maximum percentage of soil moisture content was 37.10% (0-15cm) recorded at paddy field while the minimum percentage of soil moisture content was 18.63% (0-15 cm) recorded at bamboo garden.The soil moisture content at different soil depth of study site i.e. paddy field, home garden, bamboo garden and forest came out to be 37.1%, 25.93%,18.63%,21.6% respectively of the top surface of soil while its sub surface of soil layer came out to be 36.06%, 25.4%, 22.9% and 21.66% respectively. The high moisture content in the cultivated land soil serves to protect the roots from drying. It also allows the expansion and penetration of roots into soil as well as uptake of water and nutrient from the soil (Edena et al, 2011). The minimum moisture content in bamboo garden soils could be due to the presence of hydrocarbons and polycyclic aromatic hydrocarbons, which can cause an increase in soil hydrophobicity, which leads to a decline in the water holding capacity of soil (Balks et al, 2002). It also might be due to more slope percentage which enhances the soils to runoff of water from the study area. The soil moisture content increases with the progressive increase of restored sites which may be due to vegetation over the surrounding areas. Thus, they prevent the direct exposure of soil surface and loss of soil water through evaporation. The accumulation of vegetation biomass influences the vegetation growth and soil properties in the mining area. The outcome of the successful eco-restoration usually depends on the nature of plant distribution in the restored sites (Eamus et al. 2005). Thus, the soil physical property increases with increasing age of restoration providing succession restoration in the land use system areas.

Figure 4: Soil moisture content (%) of different land use pattern in Ziro.

Bulk density

Paddy field has the highest bulk density with its values as 1.06 g/cm3 and bamboo garden has the lowest bulk density 0.9g/cm3 .The values of surface soil content of bulk density for different land use pattern i.e. paddy field, home garden, bamboo garden and forest are 1.06, 1.00, 0.9, and 1.00 respectively while for sub surface it was 0.96, 0.92, 0.94, and 0.88 respectively. The descending values of the surface content of bulk density are paddy field (1.06) > home garden(1.00)> forest(1.00)>bamboo garden(0.9).The low bulk density found in bamboo garden soils indicate that the soils tend not to be compacted and have high porosity. This property is beneficial to root activity, water infiltration into soil, and overall growth of crops. Different levels of erosion of soil depending upon the slope and management practices are also responsible for higher bulk density (Singh, G.R., and Prakash, O., 1985,). The high bulk density of paddy field is generally inversely correlated with organic matter content. Bulk density increases with increase in soil depth. Bulk density is determined by soil texture and modified by soil structure. Within any textural class a certain range in bulk density is expected and whether, within this range, bulk density is relatively low or high depends on the degree of structural development. Bulk density was not significantly affected by land use system, soil depth or their interaction effect. Bulk Density relates weight of solids to total volume of soil including solids and pores and is affected by both the nature of slids and the volume of pore. Present results are less than different workers such as Sushil kumar and Ranbir singh who worked at north-west Himalayas(2007), found the range of BD 0.96-1.54g cm-3; Bhuyan, 2012 who worked in agricultural lands of AP, found the range of Bd 0.84-1.00g cm-3; Tripathi et al., 2009.

Figure 5: Bulk density of different land use pattern in ziro.

Water holding capacity

The maximum and minimum percentage of water holding capacity of four different land use pattern i.e. paddy field, home garden ,bamboo garden and forest site came out to be 36%, 53.34%, 36%, 34.67% respectively of the top surface of the soil while sub surface layer of the soil came out to be 43.34%, 31.34%, 36%, 32.67% respectively. The highest water holding capacity was 53.34 %(0-15cm) obtained from home garden and 43.34% (15-30cm) obtained from paddy field whereas the lowest percentage was 34.67 %(0-15cm) obtained from forest and 31.34%(15-30cm) obtained from home garden. The highest water holding capacity of home garden might be because of the tightly packed soil particles. Water holding capacity had positive correlation with soil organic carbon and significant positive capacity of sub surface soil layer was found to be greater than the surface layer. This may be due to presence of low amount of humus in upper surface of the soil. Water holding capacity increases with soil depth which might be due to high amount of organic carbon and clay in the sub surface soils, which promote formation of aggregates soil particle and retention of water. WHC decreases with soil depth which might be due to high amount of organic carbon and clay in the surface than sub-surface soils, which promote formation of aggregates and retention of water (.Gupta, R.D., Arora, S., Gupta, G.D., and Sumberia, N.M., 2010).

Figure 6: water holding capacity of different land use pattern in ziro.

Particle density

Soil particle density (g / cm3) is mass of soil solids (oven-dry) per unit volume of soil solids. Particle density depends on the densities of the various constituent solids and their relative abundance. The particle density for the sub surface soil layer of different land use pattern (LUP) i.e paddy field, home garden , bamboo garden and forest was recorded as 2.63g/cm3 , 2.68g/cm3 , 2.206g/cm3 , 1.816g/cm3 Respectively while its subsurface soil layer was obtained as 2.32g/cm3, 2.97g/cm3, 2.163g/cm3 ,1.813g/cm3. Among the 4 sites home garden has the highest content of particle density at the surface of the soil but again sub surface of home garden has the highest value at the sub surface. Considering the surface and subsoil depths, the highest particle density values were recorded in the home garden cultivated land respectively, among the different land use types considered in the study. The results of the analysis of variance showed that particle density was significantly affected only by depth. The particle density under the different land uses decreased with increasing or decreasing the soil depth, which was contradictory to the findings reported by Ahmed (2002).

Table 1.soil physical properties of different land use pattern in ziro.

Land use pattern

Depth (cm)

Soil moisture

Water holding capacity

Bulk density

Particle density

Porosity

Paddy field

0-15

37.1

36

1.06

2.63

39.3

15-30

36.06

43.33

0.96

2.32

40.37

Home garden

0-15

25.93

53.33

1

2.68

36.31

15-30

25.4

41.33

0.92

2.986

29.87

Bamboo garden

0-15

18.633

36

0.9

2.206

39.9

15-30

22.9

36

0.94

2.163

42.51

Forest

0-15

21.6

34.66

1

1.816

54.2

15-30

21.66

32.66

0.88

1.813

47.61

Figure 6: Particle density of different land use pattern in ziro.

Porosity

The highest porosity (54.2) has been observed in the surface (0-15cm) of forest land (Table.1). Followed by bamboo garden (39.9), paddy field (39.3) and home garden (36.31).While Maximum soil porosity might be due to minimum tillage or other cultivation practices in that soil. However, decline in porosity leads to reduced pore size distribution. (Bhuyan et al. 2014).

The surface layer content of porosity for different land use pattern i.e paddy field, home garden, bamboo garden and forest land came out to be 39.3,36.31,39.9 and 54.2 respectively. While the sub-surface layer came out to be 40.37,29.87,42.51, and 47.61 respectively. Total soil porosity and pore size distribution were negatively affected by the intensity of land use (years of cultivation). The decline in total porosity was the result of a reduction in pore size distribution. Porosity and structure are not constant and can be altered by management, water and chemical processes.

Figure 7: Porosity of different land use pattern in ziro.

Soil Textures

The soil textures were silt loamy mostly and silt clay loamy in nature among the sites. percentage of sand in all the agricultural land i.e. the paddy field, bamboo garden, home garden and forest sites came out to be 61.333%, 78%,

74.666%, 78.333% respectively for the top surface of the soil while for the sub-surface the percentage was 54.666%, 74.666%, 78%, 74.666% respectively. The soil was dominant by silt content and very less accumulation of finer particles (clay). Soil texture was determined by Boyoucous Hydrometric method given by Allen et al. (1974). The highest amount of sand content (78.333%) (0-15cm) was recorded in at the surface layer while the lowest value (54.666%)(15-30cm) was observed in forest and paddy field site. sand content decreases with increase in soil depth except with increasing soil depth. The maximum silt content of 26.666 %(0-15cm) was observed at paddy field in the surface layer while the minimum content of silt (14.333%) was observed at forest in surface(0-15cm). The variation in silt contents might be due to the effect of erosion and runoff (Farmanullah Khan et al.,2007).in the entire study site, soil texture is dominated by silt loam with a very less amount of clay.

Variation in soil properties in different land use might be due to dynamic interactions among environmental factors such as climate, parent material, topography and land use or land cover. Higher concentration of sand might be due to the effect of a nearby stream and land filling by relatively coarser materials. Variation in soil physico-chemical properties in different fields might be due to dynamic interactions among environmental factors such as climate, parent material, topography and land cover/land use (Dengiz et. al., 2006). Clay content was highest(15%) at the sub-surface (15-30cm) soil layer of the paddy field and lowest clay content(1%) at both surface and sub-surface soil layer of the home garden. This may be due to the intensive and continuous cultivation which might cause compaction on the surface that reduces translocation of silt and clay particles within the different layers and due to mixing up by tillage activities in agreement with the findings reported by wakene (2001) and Jaiyeoba(2001). In all the land use types, the contents of sand and silt fractions increased with soil depth. However, there were no textural class differences among the four land use types. According to particle size classifications, soils of the investigation areas are generally silt loamy. The results show that clay contents are unstable. Therefore, differences in particle distribution, which can be attributed to the impact of deforestation and farming practices such as continuous tillage or cultivation and intensive grazing, can be observed. In this study, there were relatively less differences in particle size distribution among the surface layers of the forest soils under different land use types because these depths are relatively little affected by changes in land management. The results were in agreement with those reported by Sanchez et al., (1985) and these also support the assumption that the soil conditions prior to the shifts inland management were more or less similar. The spatial variability of particle size distribution plays an important role in crop production as they impact the soil texture, soil quality and soil erosion (Aderonke, and Gbadegesin, 2013). Textural class affects properties and management of soils. The smaller the particle size, the greater the total surface area of a given volume of soil. Texture also influences porosity (amount and size of pores). The loam textural class contains soils whose properties are controlled equally by clay, silt and sand separates. Such soils tend to exhibit good balance between large and small pores; thus, movement of water, air and roots is easy and water retention is adequate. Soil texture, a stable and an easily determined soil characteristic, can be estimated by feeling and manipulating a moist sample, or it can be determined accurately by laboratory analysis. Soil horizons are sometimes separated on the basis of differences in texture.

Table 2.soil texture of different land use pattern in Ziro.

Land use patterns

Depth (cm)

Soil texture

Textural class

Clay (%)

Silt (%)

Sand (%)

Paddy field

0-15

12

26.66

61.33

silt loam

15-30

15

30.33

54.66

silt clay loam

Home garden

0-15

1

21

78

silt loam

15-30

2

24.33

76.66

silt loam

Bamboo garden

0-15

17.33

8

74

silt loam

15-30

18

4.66

76

silt loam

Forest

0-15

7.33

14.33

78.33

silt loam

15-30

18

20.66

74.66

silt loam

CONCLUSION

The present study indicates that the variation in different land use systems i.e. (paddy field, home garden, bamboo garden and forest) have significant influence on physical characteristics of relatively non distributed and distributed soil system. It can be concluded that physical parameters such as soil moisture content, water holding capacity, bulk density, particle density, porosity and soil texture were highest in bamboo garden and home garden. This study revealed that Soil physical properties affect many processes in the soil that make it suitable for agricultural practices and other purposes. Texture, structure, and porosity influence the movement and retention of water, air and solutes in the soil, which subsequently affect plant growth and organism activity. Variation in the parameters infers that the most of the field land is in the stage of degradation.

The study focused on the impact of change in land cover type on soil quality inferred by the changes in physical characteristics of soil system. Adverse grazing has apparently resulted in a decrease in ground cover which probably in turn leads to a further coarseness in surface soil, loss of moisture and soil organic carbon. Most soil physical properties are associated with the colloid fraction and affect nutrient availability, biota growing conditions, and, in some cases. Biological properties in soil contribute to soil aggregation, structure and porosity, as well as soil organic matters decomposition and mineralization. Organism activity is controlled by various soil conditions and may be altered by management practices. Since many soil properties are interrelated with one another, it is difficult to draw distinct lines of division where one type of property dominates the behavior of the soil. Therefore, understanding and recognizing soil properties and their connections with one another is important for making sound decisions regarding soil use and management.

The study of different land use plays an important role in environment. From this study the particular various soil area can be studied and their physical parameters can be easily observed. If the soil is found to be unfertile then the necessary soil management can be applied for the better development of the crops or the availability of nutrients to the plants.

REFERENCES

Aber, J.D., Melillo, J.M. (1991) Terrestrial ecosystems. Saunders College Publishing, Philadelphia Adams PW, Sidle RC (1987) Soil conditions in three recent landslides in southeast Alaska. For Ecol Manag 18(2): 93–102

Anderson, J.M., and Ingram, J.S.I., 1993, Tropical Soil Biology and Fertility: A Handbook of Methods. C.A.B. International, UK.

Arunachalam, A. and k. Arunachalam (2000). Influence of gap size and soil properties on microbial biomass in a subtropical humid forest of north-east India. Plant Soil, 223: 185-193.

Basumatary A, Bordoloi PK (1992) Forms of potassium in some soils of Assam in relation to soil properties. J Indian Soc Soil Sci 40(3):443–446

Bellevue, S.A. 1992. The Soil Ecosystem. COG Organic Field crop Handbook: Ecological Agriculture projects, Mc Gill University (Macdonald Campus), Canadian Organic Growers Inc. Canada .50p.

Bhuyan.S.I, Soil nutrients status in prominent agro-ecosystems of East Siang District, Arunachal Pradesh. International Journal Of Environmental Sciences Volume 3, No 6, 2013

BinkleyD,Giardina C (1998). Why do species affect soils? The warp and woof of tree-soil interaction. Biogeochemistry 42(1–2):89–106

Black, C.A. (1965). Methods of Soil Analysis. Agronomy No. 9, Part 2 Amer. Soc. Agronomy, Madison, Wisconsin.

Black, C.A.Methods of Soil Analysis. American Society. 1965, Agron.U.S.A. (Exchangeable Calcium and magnesium).

Brady, N. C. and weil R.R. 2002. The Nature and Properties of soils. 13th ed. Prentice Hall, New Jersey, 960p.

Burke I.C., W.A. Reiners, D.S. Schimel. 1989. Organic matter turnover in a sagebrush steppe landscape. - Biogeochemistry, 7: 11-31.

Çelik, I. 2005. Land Use Effects on Organic Matter and Physical Properties of Soil in a Southern Mediterranean Highland of Turkey. Soil Till. Res., 83: 270- 277.

Cronan, C.S., Grigal, D.F., 1995. Use of calcium/aluminum ratios as indicators of stress in forest ecosystems. J. Environ. Qual. 24, 209±226.

Dimri BM, Jha MN, Gupta MK (1997) Status of soil nitrogen at different altitudes in Garhwal Himalaya. Van Vigyan 359(2):77–84

Dimri BM, Jha MN, Gupta MK (2006) Soil potassium changes at different altitudes and seasons in upper Yamuna Forests of Garhwal Himalayas. Indian For 132(5):609–614

Edema CU, Idu TE, Edema MO (2011). Remediation of soil contaminated with polycyclic aromatic hydrocarbons from crude oil. African Journal of Biotechnology, 10(7): 1146-1149

Franzmeier, D.P., Lemme, G.D. and Miles, R.J., (1985), Organic carbon in soils of north central United States, Soil science society of America journal, 49, pp 702-708.

Gairola S, Sharma CM, Ghildiyal SK, Suyal S (2012) Regeneration dynamics of dominant tree species along an altitudinal gradient in a moist temperate valley slopes of the Garhwal Himalaya.J For Res 23(1):53–63 in moist temperate valley slopes of Garhwal Himalaya, India

Geissen, U. and G.M. Guzman. 2005. Fertility of tropical soils under different land use systems- a case study of soils in Tabasco, Mexico. Applied soil Ecology 31(2006) 169-178. Available on line at www. Science direct. Com. or www. Else viey, Com/locate/ ap soil

Gong, J.L. Chen, N.FU, Y. Huang, Z. Huang and H.Peng. 2005. Effect of Land use on soil Nutrients in the Loess Hilly Area of the Loess Plateau, China. John Wiley & Sons, Ltd.

Grenon, F., Bradley, R.L. and Titus, B.D., (2004), Temperature sensitivity of mineral N transformation rates and heterotrophic nitrification: possible factors controlling the post-disturbance mineral N flush in forest floors, Soil biology and biochemistry, 36, pp 1465-1474.

Gupta MK, Sharma SD (2008) Effect of tree plantation on soil properties, profile morphology and productivity index I. Poplarin Uttarakhand. Ann For 16(2):209–224

Gupta, M.K., Jha, M.N. and Singh R.P. 1991. Organic carbon status in silver fir and spruce forest soil under different silvicultural systems. J. Indian Soc. Soil Sci. 39: 435-440.

Hodges, S.C. 1996. Soil fertility basics: N.C. certified crop advisor training. Soil Science Extension, North Carolina State University. 75p.

Jacob, J.: Rubber tree, man and environment. In: Natural rubber: Agro-management and crop processing (Ed.: P.J. George and C.K. Jacob). Rubber Board, Kottayam, India, 599-610 (2000).

Jenny, H. (1980). The Soil Resource: Origion and behavior. SpringVerlag, New York.

Jiang Y.J., Yuan D.X., Zhang C. et al. Impact of land use change on soil properties in a typical karst agricultural region of Southwest China: a case study of Xiaojing watershed, Yunan // Environmental Geology.-2006, vol.50,p.911-918

Johnston AE. Soil organic matter; effects on soil and crops soil use management; 1986:2: 97-105.

Khera N, Kumar A, Ram J, Tewari A (2001) Plant biodiversity assessment in relation to disturbances in mid-elevational forest of Central Himalaya, India. Trop Ecol 42(1):83–95

Kizilkaya R., Bayrakli B. Effects of N-enriched sewage sludge on soil enzyme activiyies // Applied Soil Ecology.-2005, vol.30,p.192-202

Lal, R., (2001), Potential of desertification control to sequester carbon and mitigate the greenhouse effect, Climate change, 51, pp 35–72.

Lal, R., 1993: Soil erosion and conservation in West Africa. P. 7-26. In D. Pimentel (eds) World soil erosion and conservation. Into Union for Conservation of Nature and Natural Resources. Switzerland.

Leskiw LA (1998) Land capability classification for forest ecosystem in the oil stands region. Alberia Environmental Protection, Edmonton

Lichon M. Human impacts on processes in karst terrenes, with special reference to Tasmania // Cave Science.- 1993,vol.20, No.2,p. 55-60

Miller R.W., R.L. Donahue. 2001. Soils in our environment. Seventh edition. Prentice Hall, Inc. Upper Saddle River, New Jersey.

Mitchell HL, Chandler RF (1939) The nitrogen nutrition and growth of certain deciduous trees of Northeastern United States. The Blackrock Forest Bulletin no. 11. Cornwall-on-the-Hudson, NY

Mortvedt J.J., 1991. Micronutrients in Agriculture. Soil Sci. Soc. Am. Inc., Madison,WI, USA.

Naiman, R.J., G. Pinay, C.A. Johnston, J. Pastor. 1994. Beaver influences on the long-term biogeochemical characteris-tics of boreal forest drainage networks. - Ecology, 75: 905–921.

Naseem, W.B. 1998. Physico-chemical characteristics of some eroded series of Rawalpindi area. M. Sc. Thesis. Barani Agric. Univ. Rawalpindi. pp. 82-84.

Nianpeng, H., Yunhai, Z., Jingzhong, D., Xingguo, H., Taogetao, B. and Guirui, Y., (2012), Land-use impact on soil carbon and nitrogen sequestration in typical steppe ecosystems, Inner Mongolia, Journal of geographical sciences, 22(5), pp 859-873.

Pandit BR, Thampan S (1988) Total concentration of P, Ca, Mg, N & C in the soils of reserve forest near Bhavnagar (Gujarat State). Indian J For 11(2):98–100

Reevas D. W. The role of organic matter in maintaining soil quality in continuous cropping systems// Soil and Tillage Research – 1997, Vol. 43, p. 131-167

Robertson GP, Vitousek PM (1981) Nitrification in primary and secondary succession. Ecology 62:376–386

Russell E.J. 1950. Soil conditions and plant growth. Biotech Books, New Delhi, India.

Santra, P., Kumawat, R.N., Mertia, R.S., Mahla, H.R. and Sinha, N.K., (2012), Spatial variation of soil organic carbon stock in atypical agricultural farm of hot arid ecosystem of India, Current science, 102,pp 1303-1309.

Sarlo, M.: Individual tree species effects on earthworm biomass in a tropical plantation in Panama. Caribbean J. Sci., 42, 419-127 (2006)

Schjønning, P., Thomsen, I.K., Moldrup, P., Christensen, B.T., (2003), Linking soil microbial activity to water-and air-phase contents and diffusivities, Soil science society of America journal, 67, pp 156–165.

Schjønning, P., Thomsen, I.K., Moldrup, P., Christensen, B.T., (2003), Linking soil microbial activity to water-and air-phase contents and diffusivities, Soil science society of America journal, 67, pp 156–165.

Sellers, P. J., Dickinson, R.E., Randall, D.A., Betts, A.K., Hall, F.G., Berry, J.A., Collatz, G.J., Denning, A.S., Mooney, H.A., Nobre, C.A., Sato, N., Field, C.B., Sellers, A.H., (1997), Modeling the exchanges of energy, water, and carbon between continents and the atmosphere, Science, 275(5299), pp 502–509.

Sims, Z.R. and Nielsen, G.A.(1986). Organic carbon in Montana soils as related to clay content and climate. Soil Science Society of America Journal 50, 1269-1271.

Singer M.J., D.N. Munns. 1991. Soils: An Introduction. Second edition. Macmillan Publishing Company, New York.

Stevenson FJ (1994) Humus chemistry, 2nd edn. Wiley, New York Thadani R, Ashton PMS (1995) Regeneration of Banj oak (Quercus leucotrichophora A. Camus) in the central Himalaya. For Ecol Manag 78:217–224

Sushil Kumar and Ranbir Singh. Erodibility Studies under Different Land Uses in North-West Himalayas. Jour. Agric. Physics, Vol. 7, pp. 31-37 (2007)

Tripathi O.P., Pandey H.N. and Tripathi R.S., (2009), Litter production, decomposition and physico-chemical properties of soil in 3 developed agroforestry systems of

Meghalaya, northeast India, African journal of plant science, 3(5), pp 139-141.

Tripathi, M. P., R. K. Panda, S. Pradhan and S. Sudhakar. (2002). Runoff Modelling of a Small watershed using satellite data and GIS. Journal of Indian Society of Remote

Woomer PL, Martin A, Albrecht A, Reseck DVS, Scharpenseel HW (1994) The importance and management of soil organic matter in the tropics. In: Woomer PL, Swift MJ (eds) The biological management of tropical soil fertility. Wiley, Chichester

1. Soil Resources and Land use systems Study in Ri-bhoi District, Meghalaya

SOIL RESOURCES AND LAND USE SYSTEMS: STUDY IN RI-BHOI DISTRICT, MEGHALAYA

L. S. Chanu, I. Laskar and S.I. Bhuyan

Department of Botany,

Pandit Deendayal Upadhyaya Adarsha Mahavidyalaya-Behali, Assam, India

Email: safibhuyan@gmail.com

Abstract

Present study was conducted in Ri-bhoi district of Meghalaya which lies between 90°20'30" E and 92°17'00" E longitude and 25°40' N to 26°20' N latitude. The soil samples were collected from four different selected Land use pattern (LUP) of the district such as forest, agricultural land (paddy field), rubber plantation and home garden. First, the detritus (grass and litter) at each sampling station were cleared from the ground surface and the soil was dug with a spade. Different parameters of soils were done with the standard methodologies. Significant variations were found among the land use systems. Land use changes from the past to the present will contribute greatly to the planning work to be done concerning the future. It will also be of help in the determination of precautions needed to be taken for the environmental sustainability, which is a concept frequently discussed nowadays.

Keywords: Conservation, Porosity, Soil fertility, Sub layer, Water holding capacity.

INTRODUCTION

Soil is a complex system where different factors such as physical, chemical and biological are held in dynamic equilibrium (Kizikaya and Bayrakli, 2005). It consists of water, minerals, air and soil organic matter. It is often classified in terms of their physical characteristic. Physical properties like texture, structure porosity and fraction of pore space, affected by composition and proportion of the soil components. On the other hand, these properties affect air and water movement in the soil and thus the soil's ability to function. The structure of the soil is the way in which the individual soil particles are bound together to form aggregates. Soil texture and texture classes are sand, slit; clay and loam denote the proportion of various particle sizes that it contains. Loams are generally the best soil for the agriculture as it contains a mixture of all particle sizes.

Soil properties of the solid phase result in both marked variability of water contents and a varying degree of resistance to the elimination of moisture. Soil with an extreme pH is strongly influenced by salts, resulting in very acid sulphated soils to highly alkaline carbonate soils. Soil acidity is enhanced by different factors such as mineral leaching, decomposition of acidic plant, industrial wastes, acid rains and certain forms of microbiology activity. However, many ions, sodium, potassium, magnesium and calcium are responsible to be alkality of the soil.

Physiochemical characteristics of soils vary in space and time which is greatly affected by variation in topography, climate, physical weathering process, vegetation cover, microbial activities and several other biotic and biotic variables. Forest cover plays a significant role in soil formation (Chapman and Reiss 1992). For example, litter fall influences physicochemical characteristics of soil such as water holding capacity, soil texture, pH of soil and nutrient availability (Johnston 1986).

The growth and reproduction of plant community cannot be understood without the knowledge of soil property. Plant growth and yield depend on the physical, chemical and biological properties of soil. Soil physical properties affect the germination and emergence of young seedling, root penetration, water movement, soil air composition and emergence of young seedling, root penetration, water movement, soil air composition and availability of plant nutrients. Soil chemical characteristics control solubility and bioavailability of essential plant nutrients and thus establish a strong relationship between soil constituents and plant productivity (Khattak, R.A. 1996). Soil and vegetation has a complex interrelationship because they develop together over again restoration of nutrients to the soil properties (Singh et al.1986). Concentration of micro elements in the soil is a good indicator of their availability to plants. Their concentration gives good information towards the knowledge of bio-chemical cycle in the soil-plant ecosystem (Pundit and Thampan, 1988). However, natural vegetation have an important effect quality than most other plant ecosystem types, through the input of litter with high lignin content, high total net primary production, high water and nutrient demand (Brinkley and Guardian 1998).

Even plant species diversity also affects soil properties as well as soil fertility (Augusto et al. 2002). The properties of soil are the important factor for the growth of the plants. The adequate theoretical and practical knowledge of various soils is therefore necessary to study.

Conversion of natural forest to other forms of land-use can provoke soil erosion and lead to a reduction in soil organic content, lost soil nutrients and modification of soil structure (Lichan, 1993; Chen et al. 2001). According to Jiang et al. (2006), soil organic matter content decreased from 38.02 g per kg in the past 20 years (1982-2003) after transformation of forest land into cultivated land. Intensive agriculture increase soil erosion and lead to change in chemical composition of the soils because of changed vegetation and altered soil water conditions. Reduction of the soil, humus content and agricultural productivity is the effect of conversion of forest.

Tree plantations are being established for different land use systems in tropical and subtropical areas, viz. shifting timber production from native forests to plantation, restoration of degraded lands, catalysts of forest succession and also as buffer zones for biodiversity conservation (Sarlo, 2006). Rubber is a deciduous plant with very fast growth and well canopy; it shows maximum litter addition to plantation floor amounting to 7 tons per hectare (Jacob, 2000). The litters are persisting on the plantation floor throughout the year and show very slow rate of decomposition. According to Zhang et al. (2007), rubber latex flow starts at 7 year age of plantation as maximum at 20 years and typically cases at 35 years. They reiterated that rubber plantation decreases soil organic carbon which is linked to latex harvesting.

Paddy soils are heterogeneous in nature. The responses from physical, chemical and biological properties along with management- induced soil changes like tillage, liming and fertilizer amendments result in soil variation within cropped fields. Apart from spatial variation, there is also temporal variation, such as nutrient status. The variation of soil properties in space and time implies that soils have varying capacity to retain and supply nutrients to rice crop. This makes it difficult to correctly manage field input applications.

A number of physico-chemical changes occur on submergence of a soil. Of these the more important are the increase in pH, increase in specific conductance and decrease in oxidation-reduction potential. The pH of an acid soil tends to increase gradually on submergence due to high concentrations of carbon dioxide content. The pH of alkali soils tends to decrease on submergence. Low soil pH promotes' the accumulation of organic acids, reduced iron and manganese. Submergence of a soil brings about an increase in the concentration of several ions like ammonium, iron and manganese in the soil solution which account for the increased specific conductance. These ions may displace bases from the exchange complex which may be ultimately lost from the soil through drainage. A high concentration of carbon dioxide would lead to an increase in specific conductance due to the formation of bicarbonate ions.

On submergence of a soil its redox potential generally decreases. Extremely reduced conditions are harmful to rice plants, especially to their roots system. Several physiological disorders of rice in ill-drained soils have been attributed to toxic concentrations of reduced products by the introduction of atmospheric oxygen into the soil.

In forest soil, the bulk density decreases with soil organic matter content (O'Neill et al. 2005). According to Cronan and Grigal (1995), the molar ratio of Calcium to Aluminums in the soil as an indicator of forest ecosystem stress resulting from acidic deposition, harvesting or other processes contribution to acid soil infertility. However, most garden soils should be slightly acidic with a pH between 6 and 7. Soil pH is amended by applying either lime or sulfur. Lime raises a soil's pH while sulfur lowers it. Prior to planting crops, in every 3-4 months lime or sulfur should be added.

Keeping above in the consideration, present study was undertaken to understand the physical and chemical properties of soil under different land use pattern which were managed through different ways by the local people. In order to achieve this objective, soil properties were analyzed at two soil depths (0-15 and 15-30 cm) under four prominent land use in RiBhoi district, Meghalaya.

REVIEW OF LITERATURE

Soil characteristics and properties are the outcome of the interplay of pedogenic factors and processes prevailing in the area. The hilly and mountainous regions are endowed with a wide range of environmental factors that exert an influence on spatial variability of soils. A review of those factors and processes that have contributed to soil formation is necessary for a better understanding of the wide range of soils.

The hill regions of India extend from Kashmir in the North-West to Arunachal Pradesh in the North-East having a wide range of macro and micro climates, physiography, geology and vegetation. Generally the climate is subtropical at altitudes less than 1,000 m, warm temperate between 1,000 and 2,000 m, cool temperate from 2,000 to 3,000m and cold temperate from 3,000 to 4,500 m. physiography of Himalayan region consists of (a) Greater Himalayas formed largely from metamorphic rocks and valleys of high altitudes (b) Lesser Himalayas composed of a variety of sendimentary rocks. A large spatial variation in soil forming factors is responsible for evolution of different soil individuals in the hill regions.

Evrendilek et al. (2004) reported that conversion of grassland into cultivated land during a 12 year period increased BD by 10.5% but decreased TP by 9.1%. Moreover, in uncultivated land, there was relatively higher organic matter (OM) making the soil loose, porous and well-aggregated that might have reduced the BD. The BD was negatively correlated with OM and increased with soil depth for all land use system. Similar, results were reported by Celik (2005).

In unprotected cultivated lands, the finer soil particles will be selectively removed by erosion, thereby increasing the proportion of the coarser particles in the soil which leaves more sand particles (Ayoubi et al., 2011) that increases particle density. In similar soil types, keeping other things constant, soil with higher proportion of sand particles have higher particle density than otherwise (Brady & Weil, 2002).

Cultivation deteriorates soil structural aggregation reducing the soil water retention capacity (Wekene, 2001). Total available water was the highest in natural forestland followed by grassland in both soil sampling depths and sites. Higher clay and OC provided large surface area required for absorption and retention of water molecules (Materechera &Akhabela, 2001). Natural forest soils have more available water holding capacity compared to the cultivated lands (Ayoubi, 2011). It is apparent, however, that soil with clayey texture has high moisture retention at field capacity and at permanent wilting point making the available water to be crops to be lower than loam textured soil (Brady & Weil, 2002).

Soil texture had a great impact on the C/N ratio (Franzmeier et al.1985; Burke et al.1989; Nianpeng et al.2012).Comparatively higher value in the surface soil layer than the subsurface might be due to high resolution and separation rates (Sakin et al.2010). The soil textures of East Siang district (Arunachal Pradesh) were sandy clay loam in nature and it was found almost similar in the entire land use pattern (Bhuyan, 2013).

The conversion of forest ecosystem to other forms of land cover may decrease the stock of OC due to changes in soil moisture and temperature regimes, and succession of plant species with differences in quality and quantity of biomass returned to the soil (Offiong & Iwara, 2012). Evrendilek et al. (2004) owed that deforestation and subsequent cultivation decreases organic matter by 48.8%. Moreover, the conversion of forest into cropland is known to deteriorate soil physical properties and making the land more susceptible to erosion since macro-aggregates are disturbed (Celik, 2005). OC is a powerful indicator for assessing soil potential productivity (Shukla et al., 2006). The relatively higher OC in soil at upper depth in all land use system could be due to relatively better return of biomass for decomposition at the surface. Wakene (2001) also reported surface soil horizons to be more biologically active in the soil systems.

The presence of dense vegetation affords the soil adequate cover thereby reducing the loss in macro and micro nutrients that are essential for plant growth and energy fluwes (iwara et al. 2011). Its status was low in the cultivated land, eucalyptus plantation at both depths and sites and grassland at 15 to 30 cm in both sites, medium total N was registered under natural forestland at Aferfida Georgis at both soil depths and at the upper depth of Abechikeli Mariam. Moreover, the N content of natural forestland at 0 to 15 cm at Abechikeli Mariam could be rated as high (Havlin et al., 1999). Ayoubi et al. (2011) reported that natural forest soil had more TN as compared to research and farmer's fields. Total nitrogen decreased consistently with depth under all land use systems corresponding to the findings of Gong et al. (2005),

Geissen and Guzman (2006) and Alemayhu (2007). Total nitrogen had positive and significant correlation with organic carbon content of the soil at both soil depths and sites. Fisseha (1992) reported highly significant correlstion (P<0.01) between nitrogen and organic matter contents of Vertisolss at Shoa Robit, Debre Zeit and Sheno sites of Ethiopia.

The available phosphorus content of all land use systems ranged from 2.61-4.91 (0-15 cm soil depth) and 1.88-3.70 mg-1 (15-30 cm). Revealing that all soils were deficient in available phosphorus according to Thomas (2000). Among the land use systems, the natural forestland contained relatively higher concentration of AP as a result of high organic matter which released phosphorus during its mineralization. Available P in all land use systems decreased with increasing soil depth. This could be due to the increased clay and reduced OM content with increasing depth of the soil. Decomposing of OM releases acids that increase the solubility of calcium phosphates (Ahan, 1993; Thompson & Troeh, 1993; Havlin et al., 1999). Available P was positively correlated with organic carbon at 0 to 15 cm soil depth and 15 to 30 cm soil depth for Abechikeli Mariam and Aferfida Georgis sites, restively. The results were in agreement with the reports of Yihenew (2002) who indicated that OC positively correlated with available P. Moreover, Berhane and Sahlemedhin (2003), Zewdu et al. (2004) and Tuma (2007) reported that available phosphorus contents of the soils were higher than that of sub soils.

Soil organic carbon can absorb and hold substantial quantities of water, up to 20 times its mass (Stevenson

1994). Soil organic carbon is responsible for building a major portion of the soil organic C pool, which regulates the soil properties viz., physical, chemical and biological properties (Woomer et al. 1994). In fine- textured soils, the content of SOM usually varies from 3 to nearly 10% by weight (Wilde, 1946). The soil C concentration varies across the landscape but more soil C variability is found at varying elevations (Powers and Schlessinger, 2002). Generally, the increase in SOM with increasing altitude is due to the addition of leaf litter annually and slow decomposition rates of organic residues under low temperature (Dimri et al. 1997). The organic C content decreased with the depth of the soil in all the forest types, which may be due to the fact that humus formation and decomposition of organic matter takes place in the upper layers (Gairola et al. 2012). According to Jha et al. (1984), if the soil is rich in SOM contents, it is definitely rich in total N contents also.

In Kashmir (Yusmarg Hill Resort), the cover, normally plants and litter, shades the soil; that is, it intercepts the cover itself instead of the soil below (Singer & Munns,1991). The presence of higher content of organic matter in the soil can be another plausible reason for lowering of the pH (Hodges, 1996). Higher levels of organic matter result in a greater number of cation exchange sites which tend to decrease the pH (Naiman et al., 1994).

The relative decline in soil pH at the soil surface of the soil under Eucalyptus plantation and natural forest land could be also due to oblong shaped canopy leading the rain to form big drops consequently enhancing leaching of basic cations as well by releasing organic acids associated with mineralization of organic matter (Mohammed et al., 2005). Soil pH increased consistently with depth that could be attributed to the downward that the increase in pH with soil depth could be associated with enhanced carbonate levels and less weathering rates.

It has been reported that forest soils should be slightly acidic for nutrient supply to be balanced (Leskiw 1998). Most of the forest types of Grahwal Himalaya had soils with pH below 6.0. This may be due to high organic matter content and the undisturbed nature of the soils in the study area (Gairola et al., 2012). Robertson and Vitousek (1981) and Adams and Sidle (1987) have also recorded low pH in undisturbed nature forests as compared to disturbed ecosystems. The reduction in pH can be attributed to accumulation and subsequent slow decomposition of organic matter, which releases acids (Hann, 1977). The values of pH recorded by Khera et al. (2001) for Kumaun Himalaya and are similar to most other studies in the Utterkhand Himalaya.

Argillic horizon in the soils of high altitude areas of Meghalaya has been reported by several workers (Nair and Chamuah 1988 and Bhattacharyya et al.1994). The soils on Cambic and on Summits have highly acidic soil (3.9 to 5.0). It is due to intense leaching leading to H+ concentration in this high rainfall region (Jenny 1980). The elevation has a significant positive correlation with organic carbon content through distinct pattern of vegetation (Sims and Nielsen 1986).

The average organic matter content of Yusmarg was found to be 7.7% in June. High organic matter content may be attributed to the rich litter deposition and due to the low mineralization caused by relatively lower temperature under the shade of dense trees and therefore to the slow rate of decomposition of organic matter (Moore, 1981). It has been reported that the decomposition rates of organic matter has a tendency to increase as weather warms and to furnish maximum plant growth conditions (Russell, 1950).

Among the land use pattern, agro-forestry system soils were less acidic which might be due to availability of greater nutrients and organic matter present in the soils. However, more acidic was recorded in the conventional farming practice (Vegetable AES) where chemical fertilizers such as urea, DAP etc. are commonly used. Different agricultural practices decrease the soil organic matter with stored carbon in soils as well as global carbon balance (Sellers et al. 1997, Lal 2001; Santra 2012). Soil organic C pool and its distribution were largely affected by management practices through various processes like quality of litter inputs, degree of decompositision, recalcitrance and turnover rate in agricultural lands (Ghani et al. 2003). Significant variation in SOC among the sites might be mainly due to the differences in the plant species composition as it was also reported by Arunachalam (2000).

The soils with high organic matter content have better supplies of organic phosphates for plant uptake than have the soils with low organic content (Miller & Donahue, 2001). Overgrazing on the plant community and soils are considered destructive because of the reduction of canopy cover, the destruction of top soil structure and compaction of soil as a result of trampling (Taylor et al., 1993; Manzano & Navar, 2000). Loss of fine fractions in soils has major influences on such properties as moisture, soil consistence, organic carbon and nutrient presence and availability (Hennessy et al., 1986). These changes in turn, influence the kind and amount of vegetation the area would support.

In moist temperate valley slopes of Garhwal Himalaya (India), nitrogen is an essential element for all growth processes in plants, especially in cold regions. If it is not available, the plant remains stunted and comparatively undeveloped (Gairola et al., 2012). Ecologists have long considered temperate forests systems to be limited by N availablility (Mitchell and Chandler 1939). Soil N is supposed to be the most limiting nutrient in a majority of ecosystems (Fenn et al. 1998). Although N is mostly present in the form of nitrates in the soil, which is very mobile and get moved freely with moisture (Gupta and Sharma 2008),the values of total N in the study area were higher in upper layers as compared to the lower layers (Gairola et al.,2012). The availability of N depends to a large extent on the amount and properties of organic matter (Hann 1977). As C and N are intimately linked and primary source of C and N is found in the soil as an organic matter, in the form of plants and animal's debris (Aber and Melillo 1991). Cultivation of leguminous crop like soybean increases the soil nitrogen through different biological processes and is affected largely by the quality and quantity of organic matter input and environmental conditions (Grenon et al. 2004). These factors are consecutively influenced by land-use and residue management (Burton et al. 2007). Constant decline in soil nitrogen and denitrification, affected by temperature, SOM, water content and oxygen content (Schjonning et al.2003, Dalal et al. 2003).

The decrease of K is caused by leaching and drainage, which results in the destruction of vegetation (Basumatary and Bordoloi 1992). Total soil K varied on different altitudes and the higher content of available K was found in surface soil in the form of exchangeable K, which may latter convert into the soil solution (Dimri et al. 2006).

The forgoing review of literature reveals that only limited studies have been conducted on soil physico-chemical properties in different land use pattern of the state in general and Ri-Bhoi district in particular. Therefore, a proper investigation is very important on soil characteristics with particular emphasis to agro-climatic conditions in Ri-Bhoi district, Megthalaya.

MATERIALS AND METHOD

Study site

Present study was conducted in Ri-bhoi district of Meghalaya which lies between 90°20'30" E and 92°17'00" E longitude and 25°40' N to 26°20' N latitude. The average elevation of Nongpoh is 485 meters (1591 feet). The total area of the district is 2448 sq. km., which is bounded in the north by the Kamrup district of Assam, east by the Karbi Anglong district of Assam, south by the East Khasi Hills and West by the West Khasi Hills district. It has 3 blocks namely Umling, Umsning and Jirang. The district headquarters is located in Nongpoh. Ri-Bhoi district is well connected with Guwahati and Shillong by National Highway 40.

Climate and rainfall

The climate of Ri-Bhoi District experience different types of climate ranging from tropical climate in the areas bordering Assam to the temperate climate adjoining the East Khasi Hills District. The areas bordering Assam experience hot-humid weather during the month of May to July of the year. In other area like Lum Roitong and Lum Sohpetbneng Plateau, the climate is severely cold during the months of winter and pleasant during summer period.

The temperature ranges from 10°C in December to 30°C in the month of July and August as recorded in Umsning Station, whereas in Byrnihat station, January and August record minimum (12.3°C) and maximum (35.2°C) temperatures respectively.

Soil condition

Soil in Ri-Bhoi district is classified into hill and plain soils. The major portion of the soil constitutes the patches of black loamy soil and lime silt. This soil is much suitable for growing both local and improved varieties of crops. The soil in the adjoining Assam also consists of heavy loams while the soil in other areas is interspersed with stones and chips. The major soil types of Ri-Bhoi district are Red loamy soil, Laterite soil and alluvial soil.

Vegetation

The vegetation of Ri-Bhoi district is tropical semi-evergreen forest. The elevation of semi-evergreen forest is 1,200m where the annual rainfall is between 1,500 to 2,000 mm. In this forest, some trees have a seasonal leafless period, with the deciduous component comprising up to one-third of the upper canopy. Clumping of evergreen and deciduous trees does occur, so some parts of the forest are dominated entirely by the deciduous trees. The number of species diversity in semi-evergreen tropical forest is generally lower than that of the evergreen zone (Tripathi, 2002).

Demography

The Bhois of Ri-Bhoi district is the sub-group of main Khasi Tribe. The majority of the Bhois speak the Bhoi dialect, although they use the Khasi dialact as a major subject in their schools. In Ri Bhoi district, there are other groups of tribe's viz. Garos, who speak the Tibeto-Burman groups of language, whereas the Karbis, Marngars,

N

Figure 2. Monthly rainfall (mm) and temperature (maximum & minimum) of the Ri-Bhoi district, Meghalaya

Mikirs, Bodos and Lalungs use Assamese as their Lingua Franca. Some speak and write Khasi too. The Bhois follow the matrilineal system. Children's bear the title of the mother and she is the safe keeper of all properties owned by her parents. As per 2011 census (provisional), Ri-Bhoi district has a total population of 2, 58,380 in 2001 census and 1, 27,312 in 1991 census with male population of 1, 32,445 and female population of 1,25,935. The rural population is 2, 33,226 and the urban population is about 25,154. The main occupation of the population in the district is agriculture.

Geology

Rare earth element (REE) minerals occur in a diverse range of igneous, sedimentary and metamorphic rocks in various geological environments such as in beach placers, per alkaline granites, syenites, pegmatites, carbonatities, residual laterites, phosphorites, hematitic granite breccias and ion adsorption clays. Several granitic plutons of alkaline and calc-alkaline nature have intruded in the Shillong plateau. This correspondence reports the significant occurrence of REEs within porphyritic and non-porphyritic varieties of cal alkaline Nongpoh pluton from Meghalaya.

The Neoproterozoic-Cambrian Nongpoh pluton is instructive into the Precambrian gneissic complex and Umsning schist belt of the Shillong plateau. The Nongpoh pluton comprises dioritic, granodioritic instructive and two granite variants, viz. porphyritic granite and non porphyritic granite. The porphyritic granite is exposed in and around Nongpoh, covering an area of about 200 sq.km in the Survey of India (So I) Top sheet No.78 O/13. Medium-grained, light to dark grey-colored non-porphyritic granite is exposed around Nongpoh.

So, the predominant geological formations of Ri Bhoi District are Archean Gneissic Complex, Shillong Group of rocks – Quartzite's, Granites and Alluvium.

Soil sampling

The soil samples were collected from four different selected Land use pattern (LUP) of the district such as forest, agricultural land (paddy field), rubber plantation and home garden. First, the detritus (grass and litter) at each sampling station were cleared from the ground surface and the soil was dug with a spade. The soil samples were collected from three locations of each LUP randomly from the two depths i.e. the upper 0 to 15 cm and the lower 15 to 30 cm.

Laboratory analysis

The samples were sealed in poly bags and were brought to the laboratory. In the laboratory the samples were air dried at room temperature, disaggregated by using a wooden mortar and sieved through a 2 mm mesh for further use and analysis. In addition, soil cores (2.6 cm diameter and 8.9 cm long) were also taken from each depth for bulk density determination.

The samples were sieved through 2mm sieve at field moist conditions and determination of soil moisture content and pH was done. Soil pH was measured in a portable digital pH meter. Air dried ground and sieved (0.25mm) samples were used for the estimation of organic C and total N content. Soil moisture content was determined by weight loss after drying 10 g of soil at 105° C for 24 hours and expressed as percentage dry weight. Colorimetric method

(Anderson and Ingram, 1993) and micro Kjeldahl distillation and titration method (Jackson, 1973) were applied to estimate organic carbon (C) and total nitrogen (N) respectively.

The texture of the soil was determined after air drying the soil samples. The air dry soil samples were passed through 2mm sieved and were subjected for texture analysis by Hygrometer method (Black, 1965) for estimation of percentage of different fractions (sand, slit and clay). The texture percentages of sand, silt, clay were plotted and categorized by using the triangular texture map.

Porosity of the soil samples was determined indirectly from the values of Bulk Density and Particle Density. The formula of Porosity is 1-(Bulk Density/Particle Density)*100.

Water Holding Capacity of soil samples were determined by filtration method. The dry clean Whatmann's filter paper was placed in a funnel and 50 g of weighted soil sample was placed inside the Whatmann's filter paper in the funnel. 50 ml of water was poured uniformly into the soil sample in funnel till a few drops of water spill out. The volume of excess water was measured and the water retained by the soil was calculated with the help of the formula, WHC= water retained by the soil/weight of the soil*100.

RESULTS AND DISCUSSION

The soil physical and chemical characteristics of the soil samples from selected land use in Ri-Bhoi District,

Meghalaya have been studied and the data were summarized in the tables and figures mentioned below.

Physical properties

Soil textures:-

The soil textures were loamy sand and sandy in nature among the sites (Table 2). Percentage of sand in the all the LUP, Rubber-Plantation, Agricultural land, Forest and Home-Garden sites came out to be 92, 90, 90 and 94 respectively for the top surface of the soil while for the sub-surface the percentage was 84, 88, 88 and 96 respectively. The texture of the soils varied from loamy sand to sand as particle size analysis revealed that the soils were found to contain more than 50 per cent of sand The soils were dominant by sand content and the accumulation of finer particle (clay and slit) were less.

The highest amount of sand content (96%) was recorded in Home-garden at the sub-surface layer while the lowest value (84%) of sand was observed in Rubber-plantation. Sand content decreases with increased in soil depth except in Home-garden. Naseem (1998) also reported that the sand content decreased with increasing soil depth.

The maximum silt content of 14 % was observed at Rubber-plantation in subsurface (15-30) while the minimum content of slit was observed at Gome-garden in subsurface (15-30). The variation in silt contents might be due to the effect of erosion and runoff (Farmanullah Khan et al., 2007).

The values of clay were observed to be same (2 %) in all different LUP.

Variation in soil properties in different land use might be due to dynamic interactions among environmental factors such as climate, parent material, topography and land use or land cover. Higher concentration of sand might be due to the effect of a nearby stream and land filling by the relatively coarser materials.

Table 1. soil physical properties of different land use pattern

Land use patterns

Properties

Depth

Rubber-plantation

Agricultural land

Forest

Home-garden

Soil moisture (%)

0-15

16.27±2.29

18.53±2.46

11.43±2.31

14.13±1.99

15-30

16.70±2.35

16.63±2.26

11.43±2.40

14.70±2.02

Water holding capacity (%)

0-15

47.33±6.39

64.00±8.92

58.67±7.92

62.67±8.85

15-30

56.00±8.12

61.33±9.01

64.00±8.98

60.00±7.96

Particle density (g cm-3)

0-15

2.03±0.27

2.84±0.38

2.21±0.32

2.39±0.34

15-30

2.35±0.34

2.42±0.35

2.02±0.27

2.47±0.35

Bulk density

(g cm³)

0-15

0.47±0.08

0.49±0.2

0.53±0.02

0.51±0.01

15-30

0.40±0.06

0.42±0.1

0.50±0.04

0.48±0.04

Porosity

0-15

76.84±0.88

82.74±1.12

76.01±1.1

78.66±0.88

15-30

82.97±0.8

82.64±1

75.24±1.12

80.56±0.9

Bulk Density

Forest soil has highest bulk density (0.53 g cm-3) and Rubber plantation (0.40) has lowest BD. This was due to more massive structure of Forest soil among other sites (Table 1). The surface soil content of bulk density for different land use pattern i.e. Rubber-Plantation, Agricultural land, Forest and Home-Garden came out to be 0.47,0.49, 0.53 and 0.51 respectively while for the sub-surface it was 0.40, 0.42, 0.50 and 0.48 respectively. In present finding, the bulk density decreases with increase in soil depth in all the different LUP sites. The descending order values of the bulk density after taking mean of all the depths is Forest (0.51) > Home-garden (0.49) > Agricultural land (0.45) >Rubber-plantation (0.43)

Present results are less than different workers such as Sushil Kumar and Ranbir Singh who work in north-west Himalayas(2007), found the range of BD 0.96- 1.54 g cm-3; Bhuyan, 2012 who worked in agricultural lands of AP, found the range of BD 0.84-1.00 g cm^3 ; Tripathi et al., 2009 who worked in 3 developed agroforestry systems of Meghalaya, Northeast India; found the range of BD 1.14 g cm-3 and 1.21 g cm-3; Gurumurthy et al. 2008 who worked on the soils under different land use systems and found the ranges from 1.48 to 1.67 mg m-3.

The low bulk density found in Rubber-plantation soils indicates that the soils are not compacted and have more porosity. This is beneficial to root activity, water infiltration into soil, and overall growth of crops. The lower bulk density implies greater pore space, improved aeration and creates a choice environment for biological activity (Werner, 1997). The high bulk density of Forest soil (uncultivated) is unexpected since the bulk density is generally inversely correlated with organic matter content. Soil with very high bulk density can limit root growth, air circulation and availability of less mobile essential plant nutrients.

Water holding capacity

Agricultural land and Forest soil have the maximum water holding capacity (64.00%) among the sites. By taking the average of both the depth for the two sites individually, it was found that the Agricultural land has more capacity to hold water. This might be due to tightly packed soil particles. And the minimum value was found in Rubber-plantation soil 47.33% (Table 1). The descending order values of the Water holding capacity after taking mean of all the depths is Agricultural land (62.66%) > Home-garden (61.34%) > Forest (61.33%) >Rubber-plantation (51.66%).Present result within the ranged reported by Bhuyan, 2012 (58.66-82.97%) in agricultural lands of Arunachal Pradesh.

Table 2. Soil texture of different land use pattern in Ri-Bhoi district.

Soil texture

Land use pattern

Depth

Rubber-plantation

Agricultural land

Forest

Home-garden

Sand (%)

0-15

92

90

90

94

15-30

84

88

88

96

0-15

2

2

2

2

Clay (%)

15-30

2

2

2

2

0-15

6

8

8

4

Slit (%)

15-30

14

10

10

2

Textural class

Loamy Sand

Loamy Sand

Loamy Sand

Sand

The amount of soil water available to plant is governed by depth of soil that roots can explore. The percentage of water holding capacity for the soil surface in the all the LUP, Rubber-Plantation, Agricultural land, Forest and Home-Garden sites came out to be 47.33, 64.00, 58.67 and 62.67 respectively, for the sub-surface the WHC percentage was 56.00, 61.33, 64.00 and 60 respectively. The surface layer (0-15) for the two sites Rubber Plantation and Forest soil have lower WHC % than the sub-surface layer (15-30). While in Agricultural land and Home-garden, the surface layer (0-15) have higher percentage of WHC than the sub-surface layer (15-30).

Porosity

In the present study porosity of the soil as shown in table 1 varies significantly between the sites. The highest porosity was recorded for Agricultural land soil 82.74 while the lowest was recorded in Forest soil 75.24 (Table1). The descending order values of the Porosity after taking mean of all the depths is Agricultural land (82.69) > Rubber-plantation (79.9) > Home-garden (79.61) > Forest (75.62). Present result is higher than the result obtained by Bhuyan et al. 2013 (61-67.7%) in Agro-ecosystem in Eastern Himalaya and Gairola et al. 2010 (46.12%) from restored mine land.

The surface layer content of porosity for different land use pattern i.e. Rubber-Plantation, Agricultural land, Forest and Home-Garden came out to be 76.84, 82.74, 76.01 and 78.66 respectively while the sub-surface layer came out to be 82.97, 82.64, 75.24 and 80.56 respectively. The two sites Agricultural and Forest soil surface layer have higher value of porosity than its sub-surface layer. While the other two sites Rubber plantation and Home-garden soil surface layer have lower value of porosity than its sub-surface layer.

Soil moisture content (SMC)

The percentage for maximum soil moisture was recorded in Agricultural land soil 18.53 while the minimum was recorded in Forest soil 11.43 (Table1). The moisture content for the surface soil of different LUP came out to be 16.27, 18.53, 11.43 and

Figure 4. Soil moisture content (%) of different land use pattern in Rib hoi district.

Figure 2. Soil water holding capacity under different land use pattern in Ri-bhoi district.

Figure 3. Particle density of soils under different land use pattern in Ri-bhoi district.

Figure 4. Bulk density of soils under different land use pattern in Ri-bhoi district.

14.13 respectively while its sub surface soil layer have 6.70, 16.63, 11.43 and 14.70 respectively. The descending order values of the Soil moisture content (SMC) after taking mean of all the depths is Agricultural land (17.58%) > Rubber-plantation (16.48%) > Home-garden (14.41%) > Forest (11.43%). In present finding, the moisture content increases with increase in soil depth in the two sites, Rubber plantation and Home-garden, while in Agricultural land the moisture content decreases with increase in soil depth. But the Forest soil has same moisture content in the two different depths.

The high moisture content of the Agricultural land soil serves to protect the roots from drying. It also allows the expansion and penetration of roots into the soil as well as uptake of water and nutrients from the soil (Edema et al., 2011). The minimum moisture content in forest soils could be due to the presence of hydrocarbons and polycyclic aromatic hydrocarbons, which can cause an increase in soil hydrophobicity, which leads to a decline in the water holding capacity of soil (Balks et al. 2002). It also might be due to more slope percentage which enhances the soils to runoff of water from the study area.

Particle density

Agricultural land has highest particle density 2.84 g cm^3 while Forest soil has lowest particle density 2.02 g cm^3 (Table 1). The Particle density content for the surface soil of different LUP came out to be 2.03, 2.84, 2.21 and 2.39 respectively while its sub surface soil layer have 2.35, 2.42, 2.02 and 2.47 respectively. The descending order values of the Particle density after taking mean of all the depths is Agricultural land (2.63 g cm^3) > Home-garden (2.43 g cm^3) > Rubber-plantation (2.19 g cm^3) > Forest (2.11 g cm^3). Among the 4 sites two sites, Rubber plantation and Home-garden have higher content of particle density at the surface of soil while the other two sites Agricultural land and Forest soil have higher value at the sub surface layer.

Chemical properties

Soil hydrogen ion concentration (pH)

Soil pH varied significantly among the sites. The hydrogen ion concentration (pH) of the 4 land use pattern (LUP) soil ranges from a minimum of Rubber-plantation soil 5.3 pH to a maximum Agricultural land 6.54 pH (Table-3) shows the variation in hydrogen ion concentration (pH) of soil. The higher values of pH which was found in Agricultural land indicated alkaline nature. However, higher soil pH in agricultural land might be due to the application of different chemical fertilizer containing sulfate and phosphate to the soil, photosynthesis or pesticides/insecticides used to increase the fertility of soil, which invariably contribute to higher pH values of soil. The available source of nutrients has also significant effect on the pH of the soil of land use patterns (Bhuyan, 2012).

Depth wise variations in soil pH were also found among the fields. In all the land use patterns except agricultural land, upper soil layer (0-15cm) showed more pH range than the sub surface layer (15-30 cm). Higher soil pH in upper soil layer might be due to high concentration of nutrients which are coming from plants parts decomposition. Minimum soil pH in rubber field might be due to less concentration of different micronutrients (Bhuyan, 2012). The surface soil pH usually increases more for urban soils than for field soils where lime is applied and agricultural management (tillage) mixes soil. Besides tillage, cropping systems influences soil properties, particularly pH. Increases in soil pH shift Fe from the exchangeable and organic forms to iron oxides fraction (Mortvedt, 1991). Leaching and runoff of soil nutrients also affect the soil pH in the fields. Present values of soil pH are lower than the values (5.47-6.67) of pH recorded by Gairola et al. 2012 from the forest composition in moist temperate valley slopes of Garhwal Himalaya, India. However, the present study is lower than the pH value (6.07-6.47) recorded by Stephen Emmauel et al. 2013.

Table 3. Soil chemical properties of different land use pattern

Land use pattern

Depth

Rubber plantation

Agricultural land

Forest

Home-garden

pH

0-15

5.38±0.12

6.24±0.22

5.9±0.08

6.48±0.28

15-30

5.3±0.12

6.54±0.08

5.7±0.07

6.18±0.22

Soil organic carbon%

0-15

1.3±0.02`

0.85±0.01

1.8±0.03

1.4±0.03

15-30

1.18±0.01

0.7±0.01

1.4±0.02

1.31±0.02

Nitrogen%

0-15

0.25±0.02

0.28±0.01

0.28±0.02

0.4±0.01

15-30

0.3±0.04

0.2±0.02

0.26±0.01

0.36±0.01

Nitrogen

Different land use patterns show the variation in total nitrogen (TN) amount in soils. Nitrogen percentages of 4 sites soil ranges from a minimum of Agricultural land soil (0.2%) to a maximum Home-garden soil (0.4%) (Table 3). The content under cultivated land was significantly lower than the other land use types (Table 3). This decline in total Nitrogen (TN) contents was observed which might be due to deforestation and subsequent cultivation practices.

Figure 5. Soil pH of different land use pattern in Rib hoi district.

Figure 6. Total nitrogen (%) of different land use pattern in Rib hoi district.

Figure 7. Soil organic carbon (%) of different land use pattern in Rib hoi district.

Nitrogen also varied significantly depth wise in different fields (Table 3). In all the sites except rubber plantation surface soil layer (0-15 cm) showed more nitrogen concentration than the sub surface layer (15-30 cm). Similar results were found by different authors in different parts of India such as Bhuyan, 2012 who worked in different land use patterns in Arunachal Pradesh; Arunachalam (2000) who worked in a subtropical humid forest of north-east India.

According to Gong et al. (2005), soil TN accumulates at the topsoil layer and TN under different land use types decline with increase in soil depth except for Rubber plantation (the sub surface layer of garden soil has higher TN content than the top surface). Different land use especially vegetation and species have different N requirements, exploiting N with varying efficiency and storing or converting N at different rates (Gong et al., 2005). The distribution of soil N is closely related to root distribution (Berger et al., 2002).

Many studies have found that nitrogen- fixing species can significantly increase soil TN while others found no correlation between the nitrogen fixing species and soil TN, accumulation in surface soil (Gong et al., 2005; Islam, 2000). However, soil nitrogen is related to soil organic matter (SOM) accumulation and indicated that the potentiality of soils for release of available nitrogen such as NH4, NO3 through mineralization. Low level of nitrogen might be due to conventional tillage practices, which initiate to rupture of mineralization of organic N substrates and also enhance the nitrous oxide (N2O) efflux. Constant decline in soil nitrogen content could be coupled through two microbial processes such as nitrification and denitrification, affected by temperature, SOM, water content and oxygen content (Schjonning et al. 2003, Dalal et al. 2003).

Present results are slightly lower than the values (0.3-.6%) obtained by Maithani et al. (1998) in north east India. However, present results are higher than the value (0.1-0.27%) obtained by Tripathi et al. (2009) from different agro-forestry systems in Meghalaya, north east India; and slightly higher than the value (0.17-0.45 %) obtained by Gairola et al. 2012 from the valley slopes of Garhwal Himalaya, India.

Soil Organic Carbon

Soil organic carbon (SOC) of the all the sites soil ranges from a minimum of Agricultural land (0.7%) to a maximum of Forest soil (1.8%) (Table 3) shows the variation in carbon of soil. All the different LUP sites have higher SOC content in the surface of soil while its value decreases with the increase in depth. The descending order values of the soil organic carbon after taking mean of all the depths is Forest (1.6%) > Home-garden (1.35%) > Rubber-plantation (1.24%) > Agricultural land (0.77%).

Variation in SOC under different LUP may be due to the differences in plant species composition and the soil fertility management (Arunachalam, 2000). Continuous cultivation processes and tillage could be other responsible factor for low SOC. Ploughing causes the breakdown of aggregates, may further increase the degradation processes by exposing organic material to biodegradation and oxidative agents (Six et al. 2000). Maximum value of SOC was recorded in Forest (Table 3). Among the land use classes, highest organic carbon contents of Forest soils might be due to higher tree density resulting in higher amount of litter addition (Kumar and Singh, 2007). Minimum SOC of Agricultural land might be due to the continuous cultivation processes and tillage (Bhuyan, 2012). Ploughing causes the breakdown of aggregates and further increases the degradation processes by exposing organic material to biodegradation and oxidative agents (Six et al., 2000). Banerjee and Badola (1980) and Gupta et al., (1991) also observed that the coniferous forests contained more organic carbon at the surface soils and decreased with profile depth. Soil organic matter plays an important role in soil physical, chemical and biological properties, which helps in creating a favorable medium for biological reactions and life support in the soil environment (Horwath, 2005). Present result is within the range of Bhuyan, 2012 (0.5-2.50%) who worked in agricultural lands of Arunachal Pradesh and lower than the result obtained by Tripathi et al. 2009 (1.33-4.49%) and Gurumurthy et al. 2008.

CONCLUSION

The present study indicates that the variations in land use systems (Rubber-plantation, Agricultural land, forest, and Home-garden) have significant influence on physico-chemical characteristic of soil. It can be concluded that physico-chemical parameters such as pH, water holding capacity (WHC), soil moisture content (SMC), porosity and particle density (PD) were found to be highest in Agricultural land (Paddy field). But the percentage of soil organic carbon (SOC) and the total nitrogen (TN) content were found to be lowest in the Agricultural land (Paddy field).

The study focused on the impact of change in land-cover type on soil quality inferred by the changes in chemical and physical properties of relatively non-disturbed and disturbed soil systems. Continuous grazing at apparently has resulted in a decrease in ground cover which probably in turn leads to a further coarseness in surface soil, loss of moisture and soil organic carbon. Variations in these parameters infer that the grassland is in the stage of degradation.

The study of different land use/ land cover has a significant role in the environment. From this study the composition of the particular soil area can be found. If the soil is found out to be unfertile then the necessary soil management can be apply for the better crop growth or the availability of nutrient to the plants.

The results of the study show that the Forest soil has high organic carbon and total nitrogen content. When this undisturbed forest land is converted into the cultivated/Agricultural land, the percentage of soil organic carbon and total nitrogen content become very less as compare to the other land use pattern. The soil quality and health were maintained relatively under the forest, whereas the influence on parameters like soil organic carbon (SOC), total nitrogen (TN) were negative on the soils of the cultivated land, suggesting the need for intervention so as to optimize and sustain the soil quality in the case of cultivated land. Special emphasis should be given for the management of soil organic matter (SOM) as many physicochemical properties are correlated with it.

Long periods of continuous cultivation of natural forest land led to changes in some of the physical and chemical properties of soils. Soil characteristics negatively affected by tillage practices are soil organic matter, aggregate stability and bulk density. Especially, after natural forest land transformation into cultivated land, decreases of organic matter have crucial effects on soil physical and chemical properties well explained the vulnerability of the structure and function of the natural forest system. Therefore, some measurements should be taken. For example, Reevas (1997) reported that conservation of tillage practices generally result in higher amount of soil organic matter, increased infiltration, increased water stable aggregates, reduced erosion and greater microbial activities when compares to conventional tillage systems.

Plants depend on the nutrients in the soil to grow and flourish. Maintaining a fertile soil is one of the most important duties of the home gardener. Soil should be tested periodically to determine pH and nutrient content. This information is used to establish a basic soil fertility level. Once this basic level is established, soil fertility is maintained by annual fertilizer application, seasonal side dressing, and organic matter maintenance. Plants depend on the nutrients in the soil to grow and flourish. Maintaining a fertile soil is one of the most important duties of the home gardener. Soil should be tested periodically to determine pH and nutrient content. This information is used to establish a basic soil fertility level. Once this basic level is established, soil fertility is maintained by annual fertilizer application, seasonal side dressing, and organic matter maintenance. A light fertilizer application at the start of the growing season helps maintain minimum soil nutrient levels. Side dressing during the growing season provides plants with the nutrients they need at critical points in their life cycle. Finally, to maintain soil structure and microbial health, organic matter should be applied at least once a year. Following these maintenance principles will ensure healthy plants and good yields for years to come.

A light fertilizer application at the start of the growing season helps maintain minimum soil nutrient levels. Side dressing during the growing season provides plants with the nutrients they need at critical points in their life cycle. Finally, to maintain soil structure and microbial health, organic matter should be applied at least once a year. Following these maintenance principles will ensure healthy plants and good yields for years to come.

Therefore, the final conclusion of the study of land use changes from the past to the present will contribute greatly to the planning work to be done concerning the future. It will also be of help in the determination of precautions needed to be taken for the environmental sustainability, which is a concept frequently discussed nowadays (Gulgun et al., 2009).

REFERENCES

Aber, J.D., Melillo, J.M. (1991) Terrestrial ecosystems. Saunders College Publishing, Philadelphia Adams PW, Sidle RC (1987) Soil conditions in three recent landslides in southeast Alaska. For Ecol Manag 18(2): 93–102

Anderson, J.M., and Ingram, J.S.I., 1993, Tropical Soil Biology and Fertility: A Handbook of Methods. C.A.B. International, UK.

Arunachalam, A. and k. Arunachalam (2000). Influence of gap size and soil properties on microbial biomass in a subtropical humid forest of north-east India. Plant Soil, 223: 185-193.

Augusto, L., Ranger, J., Binkley, D., Rothe, A., 2002. Impact of several common tree species of European temperate forests on soil fertility. Ann. For. Sci. 59, 233–253

Banerjee, S.P. and Badola, K.C. 1980. Nature and properties of some deodar (Cedrus deodara) forest soils of Chakrata forest division of U.P. Indian Foresters. 106: 558-560.

Basumatary A, Bordoloi PK (1992) Forms of potassium in some soils of Assam in relation to soil properties. J Indian Soc Soil Sci 40(3):443–446

Bellevue, S.A. 1992. The Soil Ecosystem. COG Organic Field crop Handbook: Ecological Agriculture projects, Mc Gill University (Macdonald Campus), Canadian Organic Growers Inc. Canada .50p.

Berger, T.W., C. Nevbaver, G. Glatzel. 2002. Factors controlling soil carbon and nitrogen stores in a pure stands of Norway spruce and mixed species stands in Austria.

Bhattacharyya, T., Sen, T.K., Singh, R.S., Nayak, D.C. and Sehgal, J.L.(1994). Morphology and classification of Ultisols with kandic horizon in North Eastern region. Journal of the Indian Society of Soil Science 42, 301-306.

Bhuyan.S.I, Soil nutrients status in prominent agro-ecosystems of East Siang District, Arunachal Pradesh. International Journal Of Environmental Sciences Volume 3, No 6, 2013

BinkleyD,Giardina C (1998). Why do species affect soils? The warp and woof of tree-soil interaction. Biogeochemistry 42(1–2):89–106

Black, C.A. (1965). Methods of Soil Analysis. Agronomy No. 9, Part 2 Amer. Soc. Agronomy, Madison, Wisconsin.

Black, C.A.Methods of Soil Analysis. American Society. 1965, Agron.U.S.A. (Exchangeable Calcium and magnesium).

Brady, N. C. and weil R.R. 2002. The Nature and Properties of soils. 13th ed. Prentice Hall, New Jersey, 960p.

Burke I.C., W.A. Reiners, D.S. Schimel. 1989. Organic matter turnover in a sagebrush steppe landscape. - Biogeochemistry, 7: 11-31.

Çelik, I. 2005. Land Use Effects on Organic Matter and Physical Properties of Soil in a Southern Mediterranean Highland of Turkey. Soil Till. Res., 83: 270- 277.

Champan JL, Reiss MJ. Ecology principles and application. Cambridge; Cambridge; 1992

Chen G., Gan L., Wang S.A comparative study on the microbiological characteristics of soils under different land-use conditions from karst areas of Southwest China // Chinese Journal of Geochemistry.-2001, vol.20, No.1,p.52-58

Cronan, C.S., Grigal, D.F., 1995. Use of calcium/aluminum ratios as indicators of stress in forest ecosystems. J. Environ. Qual. 24, 209±226.

Dalal, R.C., Wang, W.J., Robertson, G.P. and Parton, W.J (2003), Nitrous oxide emission from Australian agricultural lands and mitigation options: A review, Australian journal of soil research, 41(2), pp 165–195.

de Hann S (1977) Humus, its formation, its relation with the mineral part of the soil and its significance for soil productivity. In: Organic matter studies, vol 1. International Atomic Energy Agency, Vienna, pp 21–30

Dimri BM, Jha MN, Gupta MK (1997) Status of soil nitrogen at different altitudes in Garhwal Himalaya. Van Vigyan 359(2):77–84

Dimri BM, Jha MN, Gupta MK (2006) Soil potassium changes at different altitudes and seasons in upper Yamuna Forests of Garhwal Himalayas. Indian For 132(5):609–614

Edema CU, Idu TE, Edema MO (2011). Remediation of soil contaminated with polycyclic aromatic hydrocarbons from crude oil. African Journal of Biotechnology, 10(7): 1146-1149

Evrendilek, F., Çelik, I. and Kilic, S. 2004. Changes in Soil Organic Carbon and other Physical Soil Properties along Adjacent Mediterranean Forest, Grassland, and Cropland Ecosystems in Turkey. J. Arid Environ., 59: 743-752.

Farmanullah Khan, S.U. Khan, M.S. Sarir and R.A. Khattak. Effect of land leveling on some physico-chemical properties of soil in district dir lower. Sarhad J. Agric. Vol. 23, No. 1, 2007.

Fenn ME, Poth MA, Aber JD, Boron JS, Bormann BJ, Johnson DW, Lenly AD, McNulty SG, Ryan DF, Stottlemeyer R (1998) Nitrogen excess in North American ecosystems: predisposing factors, ecosystem responses and management strategies. Ecol Appl 8(3):706–733

Franzmeier, D.P., Lemme, G.D. and Miles, R.J., (1985), Organic carbon in soils of north central United States, Soil science society of America journal, 49, pp 702-708.

Gairola S, Sharma CM, Ghildiyal SK, Suyal S (2012) Regeneration dynamics of dominant tree species along an altitudinal gradient in a moist temperate valley slopes of the Garhwal Himalaya.J For Res 23(1):53–63 in moist temperate valley slopes of Garhwal Himalaya, India

Geissen, U. and G.M. Guzman. 2005. Fertility of tropical soils under different land use systems- a case study of soils in Tabasco, Mexico. Applied soil Ecology 31(2006) 169-178. Available on line at www. Science direct. Com. or www. Else viey, Com/locate/ ap soil

Ghani, A., Dexter, M. and Perrott, K.W., (2003), Hot-water extractable carbon in soils: A sensitive measurement for determining impacts of fertilisation, grazing and cultivation, Soil biology and biochemistry, 35, pp 1231-1243.

Gong, J.L. Chen, N.FU, Y. Huang, Z. Huang and H.Peng. 2005. Effect of Land use on soil Nutrients in the Loess Hilly Area of the Loess Plateau, China. John Wiley & Sons, Ltd.

Gong, J.L. Chen, N.FU, Y. Huang, Z. Huang and H.Peng. 2005. Effect of Land use on soil Nutrients in the Loess Hilly Area of the Loess Plateau, China. John Wiley & Sons, Ltd.

Grenon, F., Bradley, R.L. and Titus, B.D., (2004), Temperature sensitivity of mineral N transformation rates and heterotrophic nitrification: possible factors controlling the post-disturbance mineral N flush in forest floors, Soil biology and biochemistry, 36, pp 1465-1474.

Gulgun B., Yoruk I., Turkyilmaz B. et al. Determination of the effects of temporal change in urban and agricultural land uses as seen in the example of the town of Akhisar, using remote sensing techniques// Journal of Environmental Monitoring and Assessment.- 2009, vol. 150,p.427-436.

Gupta MK, Sharma SD (2008) Effect of tree plantation on soil properties, profile morphology and productivity index I. Poplarin Uttarakhand. Ann For 16(2):209–224

Gupta, M.K., Jha, M.N. and Singh R.P. 1991. Organic carbon status in silver fir and spruce forest soil under different silvicultural systems. J. Indian Soc. Soil Sci. 39: 435-440.

Havlin, J.L., J.D. Beaton, S.L. Tisdale and W.L. Nelson. 1999 Soil fertility and fertilizers. Prentice Hall, New Jersely. 499p.

Hodges, S.C. 1996. Soil fertility basics: N.C. certified crop advisor training. Soil Science Extension, North Carolina State University. 75p.

Horwath,W.R.(2005). The Importance of soil organic matter in the fertility of organic production systems. Western Nutrient Management Conference, 6: 244-249.

Islam, K. R. and Weil, R. R. 2000a. Land Use Effects on Soil Quality in a Tropical Forest Ecosystem of Bangladesh. Agriculture Ecosys. Environ., 79: 9-16.

Jackson, M.L., 1973. Soil chemical analysis. Printice Hall Pvt. Ltd., New Delhi.

Jacob, J.: Rubber tree, man and environment. In: Natural rubber: Agro-management and crop processing (Ed.: P.J.

George and C.K. Jacob). Rubber Board, Kottayam, India, 599-610 (2000).

Jenny, H. (1980). The Soil Resource: Origion and behavior. SpringVerlag, New York.

Jha MN, Rathore RK, Pande P (1984) Soil factor affecting the natural regeneration of silver fir and spruce in Himachal Pradesh. Indian For 110(3):293–298

Jiang Y.J., Yuan D.X., Zhang C. et al. Impact of land use change on soil properties in a typical karst agricultural region of Southwest China: a case study of Xiaojing watershed, Yunan // Environmental Geology.-2006, vol.50,p.911-918

Johnston AE. Soil organic matter; effects on soil and crops soil use management; 1986:2: 97-105.

K.T.Gurumurthy et al., 2008, Changes in physico-chemical properties of soils under different land use systems, Karnataka J. Agric. Sci., 22(5) (1107-1109) 2009

Khattak, R.A. 1996. Chemical properties of soil. IN: Soil Sci. E. Bashir and R. Bantel (eds.). National Book Foundation, Pakistan,6;167-199.

Khera N, Kumar A, Ram J, Tewari A (2001) Plant biodiversity assessment in relation to disturbances in mid-elevational forest of Central Himalaya, India. Trop Ecol 42(1):83–95

Kizilkaya R., Bayrakli B. Effects of N-enriched sewage sludge on soil enzyme activiyies // Applied Soil Ecology.-2005, vol.30,p.192-202

Lal, R., (2001), Potential of desertification control to sequester carbon and mitigate the greenhouse effect, Climate change, 51, pp 35–72.

Lal, R., 1993: Soil erosion and conservation in West Africa. P. 7-26. In D. Pimentel (eds) World soil erosion and conservation. Into Union for Conservation of Nature and Natural Resources. Switzerland.

Leskiw LA (1998) Land capability classification for forest ecosystem in the oil stands region. Alberia Environmental Protection, Edmonton

Lichon M. Human impacts on processes in karst terrenes, with special reference to Tasmania // Cave Science.-1993,vol.20, No.2,p. 55-60

Maithani, K., Arunachalam, A., Tripathi, R.S., and Pandey, H.N., 1998, Influence of leaf litter quality on N mineralization in soils of subtropical humid forest re-growths. Biology and Fertility of Soils 27: 44–50.

Manzano, M.G., J. Na´var. 2000. Processes of desertification by goats overgrazing in the Tamaulipan thornscrub (matorral) in north-eastern Mexico. - Journal of Arid Environments, 44: 1–17.

Miller R.W., R.L. Donahue. 2001. Soils in our environment. Seventh edition. Prentice Hall, Inc. Upper Saddle River, New Jersey.

Mitchell HL, Chandler RF (1939) The nitrogen nutrition and growth of certain deciduous trees of Northeastern United States. The Blackrock Forest Bulletin no. 11. Cornwall-on-the-Hudson, NY

Mohamed AG (2005). Improvement of traditional Acacia Senegal agro-forest: Ecophysiological characteristics as indicarors for tree-crop interaction on sandy soil in western Sudan. PhD dissertation. University of Helsinki, Finland.

Moore T.R. 1981. Litter decomposition in a subarctic spruce - lichen woodland, eastern Canada. - Ecology, 65: 299-304.

Mortvedt J.J., 1991. Micronutrients in Agriculture. Soil Sci. Soc. Am. Inc., Madison, WI, USA.

Mortvedt J.J., 1991. Micronutrients in Agriculture. Soil Sci. Soc. Am. Inc., Madison,WI, USA.

Naiman, R.J., G. Pinay, C.A. Johnston, J. Pastor. 1994. Beaver influences on the long-term biogeochemical characteris-tics of boreal forest drainage networks. - Ecology, 75: 905–921.

Nair, K.M. and Chamuah, G.S. (1988). Characteristics and classification of some pine forest soils of Meghalaya. Journal of the Indian society of Soil Science 36, 142-145.

Naseem, W.B. 1998. Physico-chemical characteristics of some eroded series of Rawalpindi area. M. Sc. Thesis. Barani Agric. Univ. Rawalpindi. pp. 82-84.

Nianpeng, H., Yunhai, Z., Jingzhong, D., Xingguo, H., Taogetao, B. and Guirui, Y., (2012), Land-use impact on soil carbon and nitrogen sequestration in typical steppe ecosystems, Inner Mongolia, Journal of geographical sciences, 22(5), pp 859-873.

Pandit BR, Thampan S (1988) Total concentration of P, Ca, Mg, N & C in the soils of reserve forest near Bhavnagar (Gujarat State). Indian J For 11(2):98–100

Reevas D. W. The role of organic matter in maintaining soil quality in continuous cropping systems// Soil and Tillage Research – 1997, Vol. 43, p. 131-167

Robertson GP, Vitousek PM (1981) Nitrification in primary and secondary succession. Ecology 62:376–386

Russell E.J. 1950. Soil conditions and plant growth. Biotech Books, New Delhi, India.

Saikhe, H., C. Varadachari and K.Ghosh. 1998a Changes in carbon, nitrogen and phosphorus levels due to deforestation and cultivation. A Case Study in simplipal National Park, India. Plant Soil 198:137-145.

Sakin,E., (2010), Carbon balance and atocks of Soils Southeast Anatolia Region (GAP), PhD Thesis, Graduate school of natural and applied sciences department of soil science, Harran university.

Santra, P., Kumawat, R.N., Mertia, R.S., Mahla, H.R. and Sinha, N.K., (2012), Spatial variation of soil organic carbon stock in atypical agricultural farm of hot arid ecosystem of India, Current science, 102,pp 1303-1309.

Sarlo, M.: Individual tree species effects on earthworm biomass in a tropical plantation in Panama. Caribbean J. Sci., 42, 419-127 (2006)

Schjønning, P., Thomsen, I.K., Moldrup, P., Christensen, B.T., (2003), Linking soil microbial activity to water-and air-phase contents and diffusivities, Soil science society of America journal, 67, pp 156–165.

Schjønning, P., Thomsen, I.K., Moldrup, P., Christensen, B.T., (2003), Linking soil microbial activity to water-and air-phase contents and diffusivities, Soil science society of America journal, 67, pp 156–165.

Sellers, P. J., Dickinson, R.E., Randall, D.A., Betts, A.K., Hall, F.G., Berry, J.A., Collatz, G.J., Denning, A.S., Mooney, H.A., Nobre, C.A., Sato, N., Field, C.B., Sellers, A.H., (1997), Modeling the exchanges of energy, water, and carbon between continents and the atmosphere, Science, 275(5299), pp 502–509.

Shukla, M. K., Lal, R. and Ebinger, M. 2006. Determining Soil Quality Indicators by Factor Analysis. Soil Till. Res., 87: 194-204.

Sims, Z.R. and Nielsen, G.A.(1986). Organic carbon in Montana soils as related to clay content and climate. Soil Science Society of America Journal 50, 1269-1271.

Singer M.J., D.N. Munns. 1991. Soils: An Introduction. Second edition. Macmillan Publishing Company, New York.

Singh AK, Parsad A, Singh B (1986) Availability of phosphorus and potassium and its relationship with physico-chemical properties of some forest soils of Pali-range (Shahodol, M.P.). Indian For 112(12):1094–1104

Six, J.,K. Paustian, E.T.Elliott and C.Combrink (2000). Soil structure and organic matter: I. Distribution of aggregates-size classes and aggregate-associated carbon. Soil Science Society of America Journal, 37: 509-513.

Stevenson FJ (1994) Humus chemistry, 2nd edn. Wiley, New York Thadani R, Ashton PMS (1995) Regeneration of Banj oak (Quercus leucotrichophora A. Camus) in the central Himalaya. For Ecol Manag 78:217–224

Sushil Kumar and Ranbir Singh. Erodibility Studies under Different Land Uses in North-West Himalayas. Jour. Agric. Physics, Vol. 7, pp. 31-37 (2007)

Tamirat Tsegaye, Mesfin Abebe, Tekalign Mamo. 1996. Vertisols of the central highland of Ethiopia, physical and chemical characterization and classification. pp. 46-77. In: Tekalign Mamo and Mitiku Haile (eds.). Soil- The Resource Base for Survival.

Taylor, jr. ch. a., n.e. garza, t.d. brooks. 1993. Grazing systems on the Edwards Plateau of Taxas: are they worth the trouble?. - Rangelands, 15(2): 53–57.

Tripathi O.P., Pandey H.N. and Tripathi R.S., (2009), Litter production, decomposition and physico-chemical properties of soil in 3 developed agroforestry systems of Meghalaya, northeast India, African journal of plant science, 3(5), pp 139-141.

Tripathi, M. P., R. K. Panda, S. Pradhan and S. Sudhakar. (2002). Runoff Modelling of a Small watershed using satellite data and GIS. Journal of Indian Society of Remote Sensing. 30(1&2): pp 39-52.

Werner, M.R. 1997. Soil Quality characteristics during conversion to organic orchard management. Applied soil Ecology 5:151-167.

Wilde SA (1946) Forest soils and forest growth. Publishers Periodical Experts Book Agency

Woomer PL, Martin A, Albrecht A, Reseck DVS, Scharpenseel HW (1994) The importance and management of soil organic matter in the tropics. In: Woomer PL, Swift MJ (eds) The biological management of tropical soil fertility. Wiley, Chichester

Zhang, M., X.H. Fu, W.T. Feng and X. Zou: Soil organic carbon in pure rubber and tea-rubber plantations in South-Western China. Tropical Ecology, 48, 201 – 207 (2007).

2. Plant Resources of Some Tropical Forest Stands of Ri-bhoi District, Meghalaya

PLANT RESOURCES OF SOME TROPICAL FOREST STANDS OF RI-BHOI DISTRICT, MEGHALAYA

I. Laskar and S.I. Bhuyan

Department of Botany,

Pandit Deendayal Upadhyaya Adarsha Mahavidyalaya-Behali, Assam, India

Email: safibhuyan@gmail.com

Abstract

Floristic studies also help us to understand the basic aspects of biology such as speciation, isolation, endemism and evolution. A total ten sites of tropical forest prevalent in the Byrnihut region of Ri-Bhoi district around the USTM were selected for details study. For the tree sampling 10 quadrats of 10×10m were laid randomly in all the selected study sites, similarly 10 quadrats of 5×5m were laid randomly for shrubs/saplings and 10 quadrats of 1×1m for ground vegetation. A total of 72 species were recorded in the quadrates, out of these 18 tree species (belongs to11 families) of angiosperm were recorded and similarly 30 species of shrubs (21 families) and 24 species of herbs (16 families) were recorded. Most of the plants were used by the local people. It is evident from the above study that the Ri-Bhoi district of Meghalaya is rich in plant wealth, which indicates the importance for biodiversity conservation. The frequent flood, grazing, fishing etc. are some of the factors, which is responsible for the depletion of the biodiversity. It is an urgent need to take action and create awareness about the usefulness of the flora so that people can save this wealth. Cultivation of threatened medicinal plants should be encouraged by the local community in order to relieve pressure on these plants.

Keywords: Community conservation, Forest wealth, Fragmentation, Threatened species.

INTRODUCTION

Floristic diversity refers to the variety and variability of plants in a given region. It refers to the number of types or taxa in a given region or group. Floristic diversity can be measured at any level from overall global diversity to ecosystem, community, species, populations, individuals and even to genes within a single individual. The floristic studies are considered as the backbone of the assessment of phytodiversity, conservation management and sustainable utilization of bioresearches of a region. They are helpful in providing clues of changing floristic pattern, new invasions, current status, rare, endemic and threatened (RET) taxa in a phytogeographical area. Floristic studies also help us to understand the basic aspects of biology such as speciation, isolation, endemism and evolution. A lot of ecological factors, mostly biotic, change the floristic components. The total number of species may be changed; dominant species may be replaced with other species; the floristic composition, i.e.; family: genus: species ratio may be changed. Detailed assessment of floristic diversity at all the three levels i.e. genetic, species and ecosystem (community and habitat) diversity is essential, without which it is difficult to plan or launch any policies and programmes for conservation purpose. Besides the detailed assessment of floristic diversity of area of conservation, endemic, endangered and medicinally important plant species and cause of forest destruction are equally important in assigning conservation values. Biodiversity is the characteristics of nature and is the basis for ecological stability. It refers to the variety and variability among living organization, the ecological complexes in which they occur, and the ways in which they interact with each other and their environment. At present, biodiversity is a result of a series of turnovers in the rate of evolution and extinction since the geological past. One of the greatest challenges facing society today is the need to address the unsustainable use of natural resources. In an ideal world, all biodiversity conservation needs should be addressed without jeopardizing human aspirations for social and economic development. Plant resources provide materials for survival, medicinal, forage values, but also possess and preserve cultural heritages, biological information and indigenous knowledge. India has a rich and varied heritage of biodiversity, encompassing a wide spectrum of habitats from tropical rain forests to alpine vegetation to coastal wetlands. India has 26 recognized endemic centers that are home to nearly one third of all the flowering plants identified and described till now.

Floristic diversity of a habitat contains wild species and genetic variation useful in the development of agriculture, medicines and industry. The ability of species to establish and persist in changing light environments has been a topic of considerable study, and in many habitats is a major factor in determining floristic patterns inherent on a landscape. Grime (1977) developed a general theory for describing plant survival strategies under various levels of resource stress and disturbance conditions. He designated three groups of strategists: plants that persist in zones of low stress and high disturbance, plants that persist in zones of high stress and low disturbance, and those that persist under conditions of low stress and low disturbance (competitive plants). Grime's work applies as a general model for many types of resource stress and disturbance types, and has been applied to studies conducted on light stress and forest herbs.

A flora consists of description of all the wild or cultivated plants contained in the country in question, so drawn up and arranged that the student may identify with the corresponding description any individual specimen which he may gather. The correct identification of every plant is very important, since it is the key to its literature. There are various types of flora such as native flora, agricultural flora or garden flora, weed flora, etc. The main objective of the flora is to afford the means of determining any plant growing in the area. It is usually accompanied by combinations of keys and description. A good Flora is one that provides work for correct identification of plant and their utilization could be taken on scientific and systematic basis. Since plants of the world are extremely variable. It is well known that floristic composition is determined by environmental factors.However, the composition influences biodiversity patterns at regional scales and further reflects both anthropogenic and natural disturbances. Therefore, floristic characteristics and biodiversity patterns are often influenced by environmental factors and anthropogenic disturbances. Conservation of biodiversity is essential for the proper functioning of ecosystems and for the maintenance of the environmental services they provide. Biological diversity is of fundamental importance to the functioning of all natural and human engineered ecosystems and by extension to the ecosystem. The survival of man is intimately related to the availability of different plant resources. The plant wealth of a country is its pride and acquiring knowledge of flora and vegetation is of immense scientific and commercial importance. Biodiversity provides to human kind enormous direct economic benefits an array of indirect essential; services through natural ecosystems and plays a prominent role in modulating ecosystem function and stability. The rapid loss in floristic diversity and changing pattern of vegetation due to various biotic and abiotic factors has necessitated the qualitative and quantitative assessment of vegetation. However numbers of study on community dynamics and phytogeographic affinities have been conducted qualitatively as well as quantitatively.

The ethnobotany (coined by John Harshberger in 1895) is the study that investigates complex relationship between plant and human culture. Ethnobotany was described as the study of direct interaction between human and plant population through its culture each human population classifies plants, develops attitudes and beliefs and learns the use of plants, while human behavior has a direct impact on the plant communities with which they interact, the plants themselves also impose limitation on humans, these mixture interaction are the focus of ethnobotany. Ethnobotany plays an important role in understanding the dynamic relationships between biological diversity and social and cultural systems. The plants have been used for food and medicine since the beginning of human civilization, so the history of ethnobotany is as old as human civilization. Ethnobotany is perhaps the most important method to study natural resources and their management by indigenous people. It enables us to work with local people to explore knowledge based on experiences of ages. People of Meghayala use plants for various ailments and for long time they have been dependent upon plant resources for their food, health, shelter, fuel and other purposes. The medicinal plants of Meghayala are specific and their distribution is restricted to small areas. Meghalaya is fairly gifted with a variety of climates, ecological zones and topographical regions. In Ri-Bhoi District of Meghalaya diversity of economically important plants is fairly rich since the area of Baridua is relatively less explored.

Ethnobotany by nature is a multi-disciplinary science of botany, ecology, and anthropology. The fundamental structure of ethnobotanical research is to examine the dynamic relationships between human populations, cultural values, and plants; recognizing that plants permeate materially, symbolically, and metaphorically many aspects of culture, and that nature is by no means passive to human action but interacted with each other. The investigation of the cultural values of plant species plays a significant role in modern medicine, farming, pharmaceutical and nutraceutical industrials sectors of a society. Ethnobotanical approaches are significant in highlighting locally important plant species, particularly for new crude drugs. The medicinal values of plant species, provides various vital modern drugs. About 25% of drugs originate from plants and many other drugs are synthetic analogues of compounds isolated from plants. Medicinal uses create awareness among local people for sustainable use of economically important plants and accumulate wide and dispersed knowledge about plant-people relationship. Most of the wild vegetables, fruits and medicinal plant uses are of little known or not known to all the outside world. Also, many of the known medicinal uses of plants have not been studied empirically in detailed for the active chemical compounds. On the wild edible vegetables and fruits, nutritional values of these plants need to be investigated.

Ethnobotany is not simply the study of the human use of plants, rather ethnobotany locates plants within their cultural context in particular societies and situates people within their ecological contexts. The study of ethnobotany is of great importance for the aid it gives to a proper understanding of the interrelations of all the several traits and of the whole material and intellectual culture of a people in its entirety. Ethnobotanical study not only prevents misapprehension and misrepresentation of observed facts, but is positively necessary in many instances to the correct diagnosis and explanation of ethnological facts. Ethnobotanists examine the culturally specific ways that humans perceive and classify different kind of plants, the things humans do to plant species such as destroying weeds or domesticating and planting specific kinds of food and medicinal plants, the ways in which various members of the plant world influences human cultures. Thus, it is an interdisciplinary subject which can also include elements of anthropology, human geography, biological conservation, pharmacology and nutrition.

The term ethnobotany has often been considered synonymous with either economic botany or with traditional medicine. Since prehistoric times, medicinal plants have been used virtually in all cultures as a source of medicine. The main traditional medicinal system includes Ayurveda, siddha. The rigveda dating between 3500 BC to 1800 BC is the earliest recorded information on medicinal plants. In India plants have been used for medicinal purposes since ancient time, as mentioned in Ayurveda.

The diversity in wild plant species contributes to household food society and health. The tribal communities are considered to be forest dwellers living in harmony with their environment. They gain part or all of their livelihood and food from forest. Tropical forests are major reservoir of plant diversity, as they harbor about 50% of the total species so far, with 12% area of the earth. These forests inhabit a large number of trees, shrubs, climbers, epiphytes, faunal wealth and a wealth of non-timber forest products. Herbal medicines are well accepted by the local people since generation with a notable degree of efficacy in alleviating different diseases. These herbal drugs are taken either in raw form or as aqueous extracts. The study of traditional knowledge was focused almost entirely on the applications and economic potential of plants by native people. At this time, the subject included more identification and categorizing of plants used by the primitive people. Since the beginning of human civilization people have used plant as food, medicine, fibers, dyes and various other purposes. The ethnobotany deals with past and present inter relationships between human cultures and the plants. The cultural values of the plant species play a significant role to modern medicine, farming and industrial sectors of a society.

REVIEW OF LITERATURE

A literature review is a text of a scholarly paper, which includes the current knowledge including substantive findings, as well as theoretical and methodological contribution to a particular topic. Literature review are secondary sources, and do not report new or original experimental work. Most often associated with academics-oriented literature, such journals are found in academic journals or such. Some of the literature reviews undertaken for the study entitled "Floristic Diversity and Ethnobotany" are given below. The rich botanical wealth of the Megharj range forest in particular zone Isari is being continuously over exploited for timber and non-timber forest products such as fodder, grasses, gums, grazing etc. The earlier work on floristic part of North Gujarat has been carried out Sexton & Sejweek (1918). Later on there was on gap were from 1917 onward Patel (2000), Ant, (2001), Jangid (2003); Desai, (2007). They worked in selected different area of North Gujarat.

During the field trip visit various photographs of rare plant species in Isari forest were taken. From this region they have reported 287 plant species. The sample area was Sabarkantha district situated in the North West part of Gujarat. The present work is the output of the continuous field study during the season winter 2008 to2009. Collected plant species were identified with the help of "The flora of Gujarat state" and flora of "The Presidency of Bombay". The total number of 58 Angiospermic families is belonging to164 genera and 287 species reported from this area. They have also noted the dominant species are Tectona grandis, Diospyrosmelanoxylon, Madhuka indica, Lantana camara, Acacia nilotica etc. in particular region at Isari. They have recorded 164 genera of Dicots and 12 genera of Monocots, 287 species of Dicot & 16 species of Monocots, belonging to 58 dicot & 6 monocot families.

A flora consists of description of all the wild or cultivated plants contained in the country in question, so drawn up and arranged that the student may identify with the corresponding description any individual specimen which he may gather (Hooker, 1897). The correct identification of every plant is very important, since it is the key to its literature (Steenis, 1957). There are various types of flora such as native flora, agricultural flora or garden flora, weed flora, etc. The main objective of the flora is to afford the means of determining any plant growing in the area. It is usually accompanied by combinations of keys and description. A good Flora is one that provides work for correct identification of plant and their utilization could be taken on scientific and systematic basis. Since plants of the world are extremely variable, therefore a wide range of Floras are available ranging from concise or field Flora to research Flora (Ali, 2008). The study area was Santh Saroola village of Kotli sattian Islamabad, Pakistan.

Their results were as follows, Flora: A total of 186 plants species belonging to 148 genera and 63 families were identified. Grasses were very common that were contributed by Poaceae 24 species, followed by Asteraceae with 20 species, Fabaceae 16 species, Euphorbiaceae 8 species, Lamiaceae & Solanaceae 7 species each and Brassicaceae 6 species; while rest of 56 families possessed species with the range of1-5. Most of the flora comprised annual plants and herbs were found as the common fraction with the percentage of (43.36%). It was followed by shrubs (24.48%), grasses (13.29%) and trees (9.79%), whereas, rest of plant habits were in negligible proportion.

With reference to life forms, Therophytes were found very frequent in terms of flora (45.99%), followed by Phanerophytes (29.95%), Hemicryptophytes (18.18%), Chaemophytes (4.81%) and Cryptophytes (1.07%). Most of the plants present in this area were herbaceous (89) natured, follow by shrub (38), grasses (24), trees (17) and climbers (11).The study carried out by them provides floristic account of plant species found in Santh Saroola, Kotli Sattian. Along the foothills and slopes, vegetation comprised shrubs and grasses. Since, the area receives sufficient rains therefore much of the area was occupied by annuals and grasses. This vegetation can utilize the transient water stored in the upper soil synchronic with precipitation. The upper dry layer of the surface deposits acts as a protective layer, moisture is stored in subsurface layers and the underlying sandstone provides added water storage capacity. All collected plant species taken by them are native to the area except few ones. Most of the flora was indigenous with few exceptions like Parthenium hysterophorus. This species is an exotic weed infesting a large area in the farm. This is a problematic weed that infested many countries (Williams & Grovers, 1980). No endemic species were been recorded from their study area. The dominance of both therophytes and phanaerophytes reflects a response to the harsh climate and anthropogenic pressure on the flora by human as well as animals (Qureshi et al.2011). According to their study as the area receives plentiful rainfall; therefore it could be other possible reason for aggregation of aforesaid life forms. Like other Asteraceous species, Carthamus oxycantha, Cirsium arvense, Conyza spp., Echinops echinatus, Saussurea spp., Lactuca spp., Parthenium hytserophorus, Silybum marianum, Tagetes minuta, Taraxacum officinale have minute seeds armed with hairy appendages that facilitate their dispersal through wind. Therefore, they are spreading at an alarming pace in the study area.

Findings from the survey as reported by the investigator were; Members of the family Cucurbitaceae, Papilionaceae and Asteraceae are rich in medicinal plants, The density of those plants where roots or tubers are used as medicine is low, Lack of scientific knowledge, has resulated in severe depletion of herbs and shrubs and severe biotic pressure of humans and grazing animals is detrimental to the regeneration of these plants, Scientific studies may elaborate the prospects of growing more and more medicinal plants successively, By proper management of medicinal plants remarkable improvement may be made on the earning of foreign exchange for the country, Certain industries based on the medicinal plants may be developed which will not only be economically viable but also help in the economic upliftment of the nation. On the basis of distribution of some medicinal plants it is observed that, there is a good scope for commercial exploitation of some pharmaceutically important medicinal plants.

Weed is a plant where it is not desired weeds can be defined as the plants growing in the wrong places from farmers point of view. In contrast to the cultivated plants the weed is the invader an uninvited guest in agricultural fields, Weeds are an excellent example of the successful struggle for existent. Out of 300000 plant species known in the world, about 30,000 are weeds. Weed flora of agriculture fields has large ecological amplitude so they multiply and flourish even in changed environmental condition. The invasive weed infesting the irrigated and non-irrigated agriculture fields and other ecosystems. The invasive weeds infesting the irrigated and non-irrigated agriculture fields and other ecosystems, the associations of various types of native& invasive weeds in crops and barren lands have become a serious problem today, Weeds compete with crop in which they grow for their resources like air, water, sunlight, minerals and other soil content etc. this results in less crop yield, poor quality of agricultural produce by the way of admixture and adulteration with weed seeds etc and consequent in less market value. Besides, weeds inflict allopathic effect on crop plants which are large througe their depressive root exudates. In addition weeds also being host several pathogens and other insect pests which are considered to be the natural enimies during the development of agricultural crop plants, Weeds are tremendously grow in crop fields and these problems are almost always face the every farmer but now a days these problematic, unwanted weeds can one of the major additional source of the medicinal importance of the human diet. These weeds are also used vaidyas for different formulation and maximum pharmaceutical industries to synthesis different drug from weeds. Those plants we call the unwanted weeds now in future that plant we will have been call edible food or medicinal plant and they are not going to cut and cultivate fields. Therefore, automatically increase the biodiversity of weeds and used for the welfare of human health and will be able to cure different major and minor diseases. Their study revealed in all 58 weed species were found to be used for medicinal purposes have been documented. Majority of the species used are from families. Asclpiadaceae, Solanaceae, Asteraceae, and Euphorbiaceae. In majority of the plants, preparations are made from Leaves, Underground parts; Stem bark, Fruits, Flower, Whole plant and Latex etc. Commonly found weeds were Argremone maxicana L., Boerhaavia diffusa, Bacopa monnieri (L.) Wettst, Calotropis gigantea L., Cardiospermum helicacabum L., Cassia tora L., Celosia argentea, Commelina benghalensis L., Cynadon dactylon L., Datura metal L., Euphorbia hirta L., Euphorbia geniculata, Lantana camera L., Phyllanthus amarus Schumach and Thonn., Portulaca oleracea L., Sesbania grandiflora L., Solanum virginianum L., Solanum xanthocarpum L., Tridax procumbens L., Xanthium strumarium Wild., Withania somnifera Dunal.etc. The investigator lastly suggests that generally weeds have never been given much importance. As they are species unwanted in the place and being ignore or thrown away. Present communication may give leads to the researchers about the utility aspect of the weeds.

Sharma et al. (1984) documented 3,924 species, 1,323 genera and 199 families, based on (BSI) secondary literature & herbarium specimens. Nayar and Sastry (1987-1990) made intensive study on threatened plants, he has reported 814 taxa of India, and about 10% of flowring plants are threatened. Singh (1988) published 'The Flora of Eastern Karnataka' he has surveyed eastern district of Karnataka like Bellary, Bidar, Bijapur, Chitradurga, Gulbarga, Kolar, Raichur and Tumkur districts. He has reported 1,421 species belongs to 696 genera & 140 families.

Shameem et al. (2011) was investigated the comparative assessment of edaphic factors and phytodiversity of herbaceous vegetation on seasonal basis at two different ecosystems in lower Dachigam National Park Kashmir in Himalaya. Later Datar & Laxminarashimhan (2011) madean extensive & floristic survey, collected 719 species, 90 genera, 122 families out of 719 species, 127 were recorded as endemic which are more than 18% of total angiosperm flora of the park.

Talbot (1909 & 1911) reported around 975 species belongs to the 90 families in his flora 'Forest Flora of the Bombay Presidency and Sind' (in two volumes). Bor (1960) deals with the 'Grasses of Burma, Celyon, India & Pakisthan', reported around 1,243 species belongs to 247 genera.

Satyanarayana & Shankarnarayana (1964) explored the dry areas of Karnataka and collected the plants in & around Bellary district. They published a paper on vegetation of Bellary district, Mysore state 'The Flora of Hassan district' by Saldanha & Nicholson (1976) reported about 899 genera belongs to 1,782 species. 'Flora of Chikmagalur' district by Yoganarasimhan et al. (1977) reported about 616 species belongs to 419 genera and 122 families, 'Flora of Coorg' (Kodagu) by Keshava Murthy & Yoganarasimhan (1990) reported 1,332 species belongs to 717 genera and 160 families. Blatter & McCann (1984) worked on 'The Bombay grasses' which reports 117 genera and 282 species of grasses.

Saldanha (1984 & 1996) was published 'The Flora of Karnataka' (out of six only two volumes were published) was reported 3,410 species. The Flora of Gulbarga District by Seetharam et al. (2000) reported about 600 species of flowering plants belonging to 101 families. 'Flora of Shimoga District' by Ramaswamy et al. (2001). 'Flora of Udapi District' by Bhat (2003) reported 1,242 species of flowring plants belonging to 694 genera and 171 families.

Staden, 2008, reviewed medicinal ethnobotany covering both the indigenous and colonial aspects of ethnopharmacology in South Africa. Area is characterized by extraordinary degree of endemism of its flora as well as unique cultural diversity. Study revealed that a high proportion, upto 70% of South African population practices ethnobotanical utilization in their cultural life and health care systems. A large number of plants used ethno medicinally were reported having the potential for novel drug development indicated by their phytochemistry.

Kumar, et al, 2009, carried out an ethnobotanical study of medicinal plants used by
the locals in Kishtwar, Jammu and Kashmir, India. They pr esented a list and uses of
medicinal plants distributed in the high altitude district and provided information
about 71 ethno- medicinally useful plants grown in region. Himalayan region harbours over 10,000 medicinal plant species, which have an influential role in lifestyle and livelihood of 700 million people living in the area. (Shengi, 2001). The temperature and alpine zones harbour highly valued medicinal plants (Ghimire et al., 2006). Western Himalayan region alone is a home to about 18,440 species of plants (Angiosperms, 8000 spp.; Lichens, 1159 spp. And Fungi, 6900 spp. 45% of the native of the native Western imalayan flora exhibits high medicinal value (Samanat et al., 1998).

Oza (1962), Sabnis (1966), and Bedi (1968) studied the flora and vegetation of Pavagadh. Bedi (1970) reported 465 angiosperInic plant species and few enlarged plant specics from the Ratanmahel. The Dcvgadh Bariya forest was thoroughly explored by a Chavan (1961). Bhatt (1971) surveyed Gora range of Rajpipla forest. A large number of plants were reported by Thaker (1969), from the Chhota Udepur and Kawant forest range in central Gujarat. In a series of publications they gave some intercsting plant distributional records with number of novelties for the whole state. Pate! (1973) recorded 720 angiospermic plants with a few records and endangered species in the Savli taluka.

Since publication of the first hook on ethnobotany 'An Introduction to Ethnobotany' (Faulks, 1958), literature on the subject is mushrooming all over the world. More impol1antly, Mudgal (1987) and Chandra (1990) compiled the diverse ethnobotanical literature existing in various languages of the world. In the Indian context, Jain (1991) presented information dealing with 2352 taxa belonging to 1174 genera Under 259 families in his book, 'Dictionary of Indian Folk Medicine and Ethnobotany' whereas Binu et al. (1922) gave outline of ethnobotanical research in India.

Joshi and Audichya (1981) enumerated 288 medicinal plants available in the forest of Rajpipla. Shah et al. (1981) gave an account of ethnobotuny for 133 species available in the forests of Saurashtra. Shah and Gopal (1985,1986) published notes on ethnomedicines of different tribal inhabitant. Gopal and Shah (1985) reported fourteen folk medicinal plants in Gujarat used for Curing Jaundice.

Nurani (1996) worked on ethnobotanical aspects of "Barda Hills." He found that total 62 plants belonging to 42 families were used by Rabaries for their life style. Tribal communities utilize the plant as food and vegetables 19 plants, 4 plants as fodder, 4 plants as fiber, 5 plants as avenue tree and 30 plants as medicinal purpose.

Kshirsagar ct. al. (2003) gave a brief account of ethnobotanical plant species of costal areas of South Gujarat. 57% population living along seacoast arc enumered.

Bhatt el al. (2003, 2(05) worked on plant used in abscesses from Saurashtra region. He found 94 taxa belonging to 38 ramilies are used and noted as antidiabetic.

Jadeja el al. (2004) described 24 drugs of plant originally used by tribal people of Saurashtra region for ahortion and easy delivery.

Jadcja et al. (2005) published series of publication of different aspects of ethnomedicobotany of Gujarat and Saurashtra region.

The floristic work of this region had been carried out by Hooker in 1862-1882. The first account of regional flora of northeastern region was made by Kanjilal et al., (1934-1940) in flora of Assam published in 4 volumes and described about 3500 dicotyledonous woody plants from the entire region. Their report has added to the understanding of vast diversity of flora of this region. Floristic work made by other botanists after Kanjilal like Deb, 1983, 1999 (Tripura), Haridasan & Rao, 1987 and Haridasan, 1999 (Meghalaya), Balakrishnan et al., 1983 (Jowai), Chowdhery et al., 1996 (Arunachal Pradesh), Gogoi and Borthakur (1991), Rowntree (1953), Singh and Mao (1998), Baishya, 1999, Singh (1999), Hyniewta (1999) are significant one. Chauhan et al., (1996) has reported both flowering and non-flowering categories from Namdapha, Changlang and Tirap District. Choudhery et al., (1996) in their floristic account dealt with representative angiosperms families of the region. In his floristic survey of the DDBR, Choudhury (2008) documented 1004 species of angiosperms belonging to 527 genera and 124 families of which 733 species dicots and 271 species are monocots. Almost all the floristic publication has also added the uses of many species thus playing a role in ethnobotanical data.

Jan et al., 2008, studied the herbal remedies used for gastro intestinal disorders inKaghan valley, Pakistan. His findings showed plants used for the gastrointestinal
disorders collected from Kaghan included 27 poly herbal r ecipes composed of 43 medicinal plant species.

Long & Li, 2004, studied the folklore of 66 medicinal plant species traditionally collected and used by the Red headed Yao people, Yunnan Province, China. They documented novel ethnobotinical uses of 27 plants 1st time in the literature. The ethnobotanical practices had deep roots in the community. Locals, especially elders, where observed having very firm belief over herbal recipes. The authors reported that younger generations were showing least interest in gaining that precious asset from their elders. An immediate baseline survey and documentation of the indigenous folklore was recommended.

Notable progress in the field of ethnobotanical research have been done recently in Northeastern states such as Sinha (1986, 1996) from Manipur, Chaturvedi & Jamir (2007) from Nagaland while Rao and Murti (1990) has highlighted the status of NE region. Boissya and Majumdar (1980) reported on folklore plants of Brahmaputra valley of Assam. Saklani & Jain, (1994) highlighted some facts about research scope on cross cultural comparison work and recorded 1296 species of plants, of which 472 are reported to be used among more than one ethnic group, and 824 species are unique to a single culture from northeast India. Sharma (1999), Sharma and Boissya (2000), Sharma (2004) reported cross cultural ethnobotanical work which includes 189 species of medicinal and aromatic plants used by the Nepalese and other tribes of Assam.

Shinwari et al., 2003, discussed threats to the sustainability of ethnomedicinal resources in northern Pakistan. Their results showed that due to external pressures many plant species were found endangered, rare and vulnerable. A high deforestation rate of 2% per year was recorded over the last 30 years in the area.

Shinwari & Gillani 2003, studied the distribution, sustainable harvest and utilization practices of medicinal plants in Astore valley, Northern Pakistan. They found 33 medicinally important herbs in local practice which were under pressure and recommended in in-situ preservation practices for those species.

MATERIALS AND METHOD

Study area

Ri-Bhoi district under Meghalaya state in India, Ri-Bhoi district lies between 25°15' and 26°15' latitude and 91°45' to 92°15' longitude, the district headquarters is located at Nongpoh, it occupies an area of 2448 km2 and has a population of 1,92,795(as of 2011), making it one of the least populous state of Meghalaya. The district East by Jaintia hills and Karbi Anglong district of Assam and on the West by West Khasi hills. The district was upgraded from sub divisional level to a full-fledged district on 4 June 1992 it was carved out from East khasi Hills. The Bhois of the Ri-Bhoi district are the sub-group of khasi tribes and majority of the people speak the Bhoi dialects there are also other tribes namely Garo, Karbis, Marngars, Mikhir, Bodo and Lalung. For the tree sampling 10 quadrats of 10×10m were laid randomly in all the selected study sites, similarly 10 quadrats of 5×5m were laid randomly for shrubs/saplings and 10 quadrats of 1×1m for ground vegetation.The studied 10 areas(quadrats) lies in area 1 (Altitude-243m, N-26°06.015', E-091°50.794'), area 2 (Altitude-239m, N-26°05.831'E-091°50.727'), area 3 (Altitude-252m, N-26°05.908', E-091°50.784'), area 4 (Altitude-225m, N-26°05.841', E-091°50.812'), area 5 (Altitude-256m, N-26°06.125', E-091°50.615'), area 6 (Altitude-212m, N-26°06.224', E-091°50.813'), area 7 (Altitude-223, N- 26°06.213', E- 091°50.834'), area 8 (Altitude-212m, N-26°05.926', E- 091°51.057'), area 9 (Altitude-221m, N-26°05.854', E-091°51.169'), and area 10 (Altitude-215m, N-26°05.872, E-091°51.147) which are located in the campus of University of Science and Technology, Meghalaya and surroundings and some part from forest area and some part from kling road and villages.

Map

Figure 1: Map of study site(Ri Bhoi district)

Geographical features

The district is characterised by rugged and irregular land surface, it includes a series of hill ranges which gradually sloped towards the north and finally joins the Brahmaputra valley. The important river flowing through this region includes the Umtrew, Umsiang, Umram andUmiam rivers

Rainfall and climate

The climate of the district is largely controlled by the South-West monsoon and seasonal winds. The Ri-Bhoi district experiences a high temperature for most part of the year being lower in altitude to the rest of Meghalaya. The average rainfall is 330 cms where two-thirds occur during the monsoon, winter being practically dry. The average rainfall of Ri-Bhoi district is 2548 mm. the maximum temperature in summer is 33.98 deg C and minimum temperature in winter is 17.50 deg C. The district has mostly tropical dense mixed forest and a small patch of temperature forest.

Demograpgy

The total population of Ri-Bhoi District is 1,92,795 as per census 2001. The Bhois of Ri-Bhoi District are the sub-group of the main Khasi Tribe. The majority of the Bhois speak the Bhoi dialect, although they use the Khasi dialect as a major subject in their schools. In Ri-Bhoi District, there are other groups of tribes viz, Garo, who speak the Tibeto – Burman groups of language, whereas the Karbis, Marngars, Mikirs, Bodos and Lalungs use Assamese as their Lingua Franca. Some speak and write Khasi too. The Bhois follow the matrilenial system. Children bear the title of the mother and she is the safe keeper of all properties owned by her parents.

Economic use Classification, Preservation and Identification

Plants were classified on the basis of their economic value, medicinal, fodder, fuel, ornamental and marketing. The plant specimens were collected and identified. The study identified medicinal plants used by the people for the treatment of some common diseases such as fever, cold, dysentery, snake bites, sores, wounds, body pain, stomach ache, head ache, boil, swellings, asthma, cough, ulcers etc.

Some of the common diseases are cured by the use of specific plants; diseases like fever and cold are cured by the plant Myricanagi, Kyllingabrevifolia, Ficuslutea, Elephantopusscaber, Celerodendrumcolebrokianum, celerodendruminfortunatum, Begonia obliqua. Swellings, boils and ulcers can be cured by Artocarpusheterophyllus, Bauhinia purpurea, Cryptocaryaandersonii, Dilleniaindica, Dioscoreabulbifera, Eupatorium odoratum, Ficuscarica, Lygodiummicrophyllum, Setariaviridis, Smilax aspera, Smilax zeylanica. Stomach ache, headache and dysentery can be treated by the plant Verbena officinalis, Tetraceravolubilis, Pouzolziamixta, Phyllanthus mirabilis, Phyllanthusniruri, Melastoma malabathricum, Osmundacinnamomea, Ficuslutea, Croton megalocarpus, and Cyperus compactus. Many more common diseases can be treated using plant species such as Artemisia indica to treat diarrhoea by drinking the juice of the plant. Dog bites and snake bites can be treated using some plant species such as Costus pulverulentus and Lantana camara. Leaf juice of Floscopascandens can be used when treating sore eyes. Roots of Dryopteris filix-mas can be used in treating uterine bleeding. Leaves and barks of Croton insularis are used in the form of tea to treat diabetes. Many more plants are used as a medicine by the local people in treating many common diseases. The district of Ri-Bhoi has many species of economically important plants but their sustainable use still remained untrapped as it is not yet properly explored.

Majority of the plants are useful in different aspect of life of the common people.

The plants were identified with their common name, photographed and sample specimens were collected from the area. In the enumeration, data were tabulated and arranged in the sequence of serial number, botanical name, family, common name habit and uses of the plant. The plant specimen was studied and identified with the help of various floras and published literature.

The varied topography, moderate rainfall and favourable climatic conditions are responsible for the high species diversity in the area. The area harbours some medicinally important plants as of which large number of medicinal plants isused in the traditional medicinal system of Ri-Bhoi district. The knowledge of herbal remedies for the treatment of various diseases rests with the traditional healers, which belong to a family of indigenous practitioners and skills have been passed on from one generation to the other only by word of mouth. Each village has one or two traditional healers or Nongaidawaikynbat as they are called locally. It was also observed that for majority of cases a single plant is administered singly but for a good number of diseases also the recipe includes a combination of many plants and plant parts. A single herbal recipe is effective for treatment of a number of ailments, which shows that a single plant is used for more than one ailment. These herbal medicines are either taken in raw form or as aqueous extracts.

RESULTS AND DISCUSSION

A total of 72 species were recorded in the quadrates, out of these 18 tree species (belongs to11 families) of angiosperm were recorded and similarly 30 species of shrubs (21 families) and 24 species of herbs (16 families) were recorded. On the basis of the habit, the plants have been grouped into trees, shrubs and herbs. Shrubs are more dominant with 30 species, followed by herbs 20 species and trees with 18 species (Figure 2). Dominant tree species in the study area comprised of, Bauhinia purpurea, Cinchona pubescens, Dendrocalamus strictus, Dillenia indica, ficuscitrifolia, Mesuaferrea, Schimawallichi, and Artocarpus heterophylla, the dominant shrub species comprised of Brugsmansia versicolor, Clerodendrum colebrookianum, Coccinia grandis, Costus speciosus, Dryopteris felix-mas, Eupatorium odoratum, Lantana camara, Osmunda cinnamomea, and Smilax zeylanica and the dominant herbs comprised of Artemisia indica, Begonia obliqua, Elephantopus scaber, Dioscorea bulbifera, Kyllinga brevifolia, Lygodium japonicum, oxalis corniculata, Phyllantusamarus, Phyllantusniruri, and verbena officinalis. The most dominant family in the study site is Poaceae followed by Moraceae which is followed by Phyllanthaceae, then Asteraceae, Cyperaceae, Fabaceae and Lamiaceae respectively (Figure 3).

The importance of conservation and utilization of plant resources in natural environment has long been recognized. Since the very beginning of human civilization, human have always been dependent on plant wealth for their major needs.Some of the plants were used by local people in many different ways in the study area. The principle uses were medicinal, cultural/religious, food, timber and other household purposes. Through most of the plants are useful to mention in one way or other.Local inhabitants living in Ri-Bhoi District of Meghalaya, particularly in forest areas use a number of wild plants and their parts for curing various diseases. The following species are medicinally important whose curative properties have been well established, of these, some plant species were used for medicinal purposes to cure various disorders which ranged from diabetes, ear infections, antiseptics, joint pains, digestive problems, snake bites, headache, stomachache, boils and sores (Table 1). Some medicinal plants areKyllingabrevifolia, Ficuslutea, Elephantopusscaber, Celerodendrumcolebrokianum, celerodendruminfortunatum, Begonia oblique, Bauhinia purpurea, Cryptocaryaandersonii, Dilleniaindica, Dioscoreabulbifera, Eupatorium odoratum, Setariaviridis, Smilax aspera, Smilax zeylanica, Phyllanthus mirabilis, Melastomamalabathricum, Osmundacinnamomea, Ficus lutea, Croton megalocarpus, Cassia fistula, Cassia siamen, Cinchona pubescens, Croton insularis.

A number of plants provide fruits and other parts, which are eaten raw, cooked or pickled viz.,Artocarpusheterophyllus, Dendrocalamushamiltonii, Dendrocalamusstrictus, Phyllanthusemblica, Morusalba.

Ornamental plants can be used to make many useful thing, some of the ornamental plants are Lagettalagetto, Poatrivialis and Dendrocalamusstrictus.

A total number of 72 species under 38 families with 56 genera were collected from the Angiosperms (Table 2). Among the Angiosperms, Dicotyledons comprises 50 species under 40 genera and 30 families and the

Monocotyledons comprise 22 species belonging to 16 genera and 9 families (Figure 4).

Hence, proper documentation and preservation of traditional skills and technology of medicinal plants is a vital necessity. Further investigation on pharmacological importance of these plants and their diversity may add new knowledge to the traditional medicinal and cultural system.

During the last few decades there has been an increasing interest in the study of medicinal plants and their traditional use in different parts of India and there are some reports on the use of plants in traditional healing by either tribal people or indigenous communities of Ri-Bhoi district. The tribes of Ri-Bhoi District of Meghalaya use many plant species for the treatment of some common diseases. The common diseases treated by the herbal practitioner were asthma, digestive problems, paralyzes, skin diseases, snake bites, dog bites, fever, cold, headache, stomachache, diarrhea, uterine problems, boils, skin infections, wounds, sores and diabetes.

Local communities depend on farming and livestock rearing. Some of the species were identified as fodder species in the area. In summer season livestock graze upper land, and in winter season the livestock is kept under the sheds or even within the houses due to heavy rainfall in the area. Fodder plants are a very important source for animals as they eat the plant to survive and reproduce. Present study includes plant species which are used medicinally by using different parts of the plant. Leaves, roots, barks and sometimes even the whole plant is crushed to make medicine which cure diseases like fever, cold, boils, skin infections etc. Local healers also have some basic knowledge on codified systems of medicine such as Ayurveda and Unami, while majority of them practices the traditional. Local healers in the study area belong to almost all castes and religion. Documenting the pattern of species diversity and their distribution creates a valuable database, useful for implementing better management and conservation of tropical forests. Presently, many forests sites of Ri-Bhoi District are subjected to various anthropogenic pressures. Data of plant diversity, such as presented in the current study on trees, shrubs and herbs will be useful in highlighting the importance of these forests for species conservation and forest management.

The present study revealed the ethnobotanical knowledge of people in Ri-Bhoi district, Meghalaya. Inthe following, medicinal details of identified plants, followed by localname, family name, part used and medicinalproperties are mentioned. Among different plant parts used by the people, the leaves are used mostfrequently to cure wounds and they appliedmostly on the external surface of the body.Generally fresh part of the plant can be used forthe preparation of medicine.When it is not in available condition, the driedleaves or roots are also used. The result of thisinvestigation showed that these local people of Ri-Bhoi district still depend on medicinal uses ofplants for the treatment of anthelmentic,alexeteric, cough, dysentery, dyspepsia, eczema,eye disease, fever, glaucoma, gonnorrhoea,headache, high blood pressure, insomnia,intestinal infections, jaundice, low bloodpressure, liver disorders, menstrual disorders,mouthwash, nervous disorders, purify blood,pain, piles, rheumatism, respiratory tract, skindisease, snake bites, small pox, whooping cough,urinary tract and many types of diseases.

Table 1. List of plants economic utility of the present study

Species

Family

Types of Uses

Uses/parts used

Acacia concinna

Fabaceae

Medicinal

Leaves and fruits are used. Its dried pods are powdered

Adiantumcapillus-veneris

Pteridaceae

Medicinal

Infusion of the plant used as lotion for centipede stings

Adiantumcunninghamii

Pteridaceae

Medicinal

Whole plant is dried and used as medicine

Alocasiaindica

Araceae

Medicinal

Juice of the leaves is used as medicine

Ardisiahumilis

Primulaceae

Medicinal

Parts like leaves and fruits are used as medicine

Artemisia indica

Asteraceae

Medicinal

Juice of the plant is used in diarrhea

Artocarpusheterophyllus

Moraceae

Medicinal

Leaves used in boils, leaves is burned and ashes used in ulcers

Bauhinia purpurea

Leguminoceae

Medicinal

Decoction prepared from plant is used as medicine for ulcers

Begonia obliqua

Begoniaceae

Medicinal

Whole plant is used as medicine for treating cold

Brugmansia versicolor

Solanaceae

Medicinal

Leaves are directly applied in the skin for treating pains, aches etc.

Calliandra calothyrsus

Fabaceae

Fodder

Leaves and shoots provide protein to animals

Carexcruciata

Cyperaceae

Medicinal

Infusion of the plant used in the treatment of catarrhs

Carexrepanda

Cyperaceae

Fodder

Used as food for livestock

Cassia fistula

Fabaceae

Medicinal

Decoction of leaves used in treating skin diseases

Cassia siamea

Fabaceae

Medicinal

Plant decoction used in treating urinary problems

Cinchona pubescens

Rubiaceae

Medicinal

Quinine extracted from bark is used in treating malaria

Clerodendrum colebrokianum

Lamiaceae

Medicinal

Leaves and roots used for cough, skin disease etc.

Clerodendrum infortunatum

Lamiaceae

Medicinal

Leaves used as remedy for fever

Cocciniagrandis

Cucurbitaceae

Medicinal

Fruits used to treat leprosy

Costus pulverulentus

Costaceae

Medicinal

Berries crushed to treat dog bites

Costus speciosus

Costaceae

Medicinal

Rhizomes used in skin diseases

Croton insularis

Euphorbiaceae

Medicinal

Leaves and bark used in the form of tea for treating diabetes

Croton megalocarpus

Euphorbiaceae

Medicinal

Barks crushed and boiled and drunk to cure stomachache

Cryptocarya andersonii

Lauraceae

Medicinal

Leaves crushed used in treating wounds

Cyperus compactus

Cyperaceae

Medicinal

Juice of the plant is used to treat stomachache

Dendrocalamus hamiltonii

Poaceae

Edible

Young shoots are eaten as vegetables

Dendrocalamus strictus

Poaceae

Raw material

Plant used in making mats, rafts, baskets etc

Dilleniaindica

Dilleniaceae

Medicinal

Barks used as medicine for sores

Dioscorea bulbifera

Dioscoreaceae

Medicinal

Dried powdered tubers are used in treating ulcers

Dryopterisfilix-mas

Dryopteridaceae

Medicinal

Roots taken internally to treat uterine bleeding

Elephantopus scaber

Asteraceae

Medicinal

Different parts of the plant used as astringent agent, used in fever

Eupatorium odoratum

Asteraceae

Medicinal

Leaves crushed to treat skin wounds

Ficus carica

Moraceae

Medicinal

Fruit used in herbal medicine for treating skin viral infections

Ficus citrifolia

Moraceae

Medicinal

An extract of the plant may have therapeutic value for chemotheraphy patients

Ficus lutea

Moraceae

Medicinal

Decoction of barks is used to treat stomach disorders, colds

Floscopas candens

Commelinaceae

Medicinal

Leaf juice used in treating sore eyes

Galpiniatransvaalica

Lythraceae

Medicinal

Fruits and roots are boiled and used as emetic

Heliotropium indicum

Boraginaceae

Medicinal

Decoction of the whole plant is used to treat thrush, diabetes

Jacquemontia paniculata

Convolvulaceae

Fodder

Provide food for birds and moths

Kyllinga brevifolia

Cyperaceae

Medicinal

Decoction of the whole plant is used as treatment for malaria, colds etc

Lagetta lagetto

Thymelaeaceae

Ornamental

Inner bark is drawn out and many ornamental things are made from it

Lantana camara

Verbenaceae

Medicinal

Crushed leaves is effective in snake bites

Lygodium japonicum

Lygodiaceae

Medicinal

Decoction of vegetative parts and spores is used as a diuretic

Lygodium microphyllum

Lygodiaceae

Medicinal

Locally in folk medicine to treat skin aliments, swellings

Mahonia aquifolium

Berberidaceae

Medicinal

Certain extracts are used in skin diseases such as psoriasis

Melastoma malabathricum

Melastomataceae

Medicinal

Decoction of leaves is used in treating stomach ache, indigestion

Mesua ferrea

Calophyllaceae

Medicinal

Seed oil is used in treating gastric, affected joints

Morus alba

Moraceae

Edible

Fruits are eaten raw and also used to make wine

Myricanagi

Myricaceae

Medicinal

Traditionally used for the treatment of liver diseases, fever

Osmunda cinnamomea

Osmundaceae

Medicinal

Decoction of roots is rubbed into affected joints as treatment, used in headache

Oxalis corniculata

Oxalidaceae

Edible

Plant is edible and the flowers is used to make tangy drinks

Panicum virgatum

Poaceae

Fodder

Used as hay for livestock

Paspalumdilatatum

Poaceae

Fodder

Provide food for animals

Pennisetum purpureum

Poaceae

Medicinal

Plant extracts are used as a diuretic and also used as herbal remedies

Phyllanthus amarus

Phyllanthaceae

Medicinal

The whole plant is used in gonorrhea, menorrhagia

Phyllanthus emblica

Phyllanthaceae

Edible

Fruits are eaten raw

Phyllanthus mirabilis

Phyllanthaceae

Medicinal

Infusion from young shoots are used as treatment for dysentery

Phyllanthus niruri

Phyllanthaceae

Medicinal

Juices from the plant is used in stomach ache, urinary disorders

Pieris japonica

Ericaceae

Other

Plant is used as pesticide and parasiticide

Poatrivialis

Poaceae

Ornamental

Used as an ornamental plant

Pouzolziamixta

Urticaceae

Medicinal

Root extract is taken for the treatment of diarrhoea and dysentery

Schimawallichii

Theaceae

Medicinal

Bark used as antiseptic for wounds, sap from stem used in ear infection

Senecio vulgaris

Asteraceae

Medicinal

The whole herb is used as medicine for diuretic, purgative

Setariasphacelata

Poaceae

Fodder

Seeds are important food source for birds

Setariaviridis

Poaceae

Medicinal

Plant is crushed and mixed with water then used in treating wounds

Smilax aspera

Smilacaceae

Medicinal

Fruit is squeezed and applied to the skin in the treatment of scabies

Smilax zeylanica

Smilacaceae

Medicinal

Root decoction used for boils, swellings

Solanum dulcamara

Solanaceae

Medicinal

Bark of the root, twigs is used in diuretic, emetic

Tetraceravolubilis

Dilleniaceae

Medicinal

Decoction of the leaves is used in treating dysentery

Trichosanthes tricuspidata

Cucurbitaceae

Medicinal

Fruits are used in curing otitis, asthma

Verbena officinalis

Verbenaceae

Medicinal

Root is astringent, used in the treatment of dysentery

Table 2. List of plants, common names with their habits found in the present in the study area

Species

Common name

Family

Habit

Acacia concinna

Shikakai

Fabaceae

Shrubs

Adiantum capillus-veneris

Venus hair fern

Pteridaceae

Herbs

Adiantum cunninghamii

Madian hair

Pteridaceae

Herbs

Alocasia indica

Giant taro

Araceae

Herbs

Ardisia humilis

Kada

Primulaceae

Herbs

Artemisia indica

Mugwort

Asteraceae

Herbs

Artocarpus heterophyllus

Jack fruit

Moraceae

Tree

Bauhinia purpurea

Kota

Leguminoceae

Tree

Begonia oblique

Lotus leaf begonia

Begoniaceae

Herbs

Brugmansiaversicolor

Angel's trumpets

Solanaceae

Shrubs

Calliandracalothyrsus

Red calliandra

Fabaceae

Shrubs

Carexcruciate

Lake sedge

Cyperaceae

Shrubs

Carexrepanda

Hay

Cyperaceae

Shrubs

Cassia fistula

Amaltas

Fabaceae

Tree

Cassia siamea

Seemia

Fabaceae

Tree

Cinchona pubescens

Fever tree

Rubiaceae

Tree

Clerodendrum colebrokianum

Indian glory bower

Lamiaceae

Shrubs

Clerodendrum infortunatum

Hill glory bower

Lamiaceae

Shrubs

Coccinia grandis

Tendli

Cucurbitaceae

Shrubs

Costus pulverulentus

Spiral gingers

Costaceae

Shrubs

Costus speciosus

Keu

Costaceae

Shrubs

Croton insularis

Sacaca

Euphorbiaceae

Tree

Croton megalocarpus

Rush foil

Euphorbiaceae

Tree

Cryptocaryaandersonii

Mountain laurels

Lauraceae

Tree

Cyperuscompactus

Umbrella sedge

Cyperaceae

Herbs

Dendrocalamushamiltonii

Male bamboo

Poaceae

Tree

Dendrocalamusstrictus

Tama bamboo

Poaceae

Tree

Dilleniaindica

Chalta

Dilleniaceae

Tree

Dioscoreabalcanica

Wild yam

Dioscoreaceae

Herbs

Dioscorea bulbifera

Air potato

Dioscoreaceae

Herbs

Dryopteris filix-mas

Male fern

Dryopteridaceae

Shrubs

Elephantopusscaber

Elephant's foot

Asteraceae

Herbs

Eupatorium odoratum

Floss flower

Asteraceae

Shrubs

Ficus carica

Fig

Moraceae

Tree

Ficus citrifolia

Shortleaf fig

Moraceae

Tree

Ficus lutea

Dahomey rubber tree

Moraceae

Tree

Floscopas candens

Green mini

Commelinaceae

Shrubs

Galpiniatransvaalica

Wild pride of India

Lythraceae

Shrubs

Heliotropiumindicum

Indian heliotrope

Boraginaceae

Shrubs

Jacquemontia paniculata

Skyblueclustervine

Convolvulaceae

Herbs

Kyllingabrevifolia

Shortleaf spikesedge

Cyperaceae

Herbs

Lagettalagetto

Lace bark

Thymelaeaceae

Shrubs

Lantana camara

Wild sage

Verbenaceae

Shrubs

Lygodium japonicum

Climbing fern

Lygodiaceae

Herbs

Lygodium microphyllum

Climbing fern

Lygodiaceae

Herbs

Mahonia aquifolium

Oregon grape

Berberidaceae

Shrubs

Melastomamalabathricum

Indian rhododentron

Melastomataceae

Herbs

Mesuaferrea

Iron wood tree

Calophyllaceae

Tree

Morus alba

White mulberry

Moraceae

Tree

Myricanagi

Box berry

Myricaceae

Herbs

Osmundacinnamomea

Cinnamon fern

Osmundaceae

Shrubs

Oxalis corniculata

Creeping oxalis

Oxalidaceae

Herbs

Panicumvirgatum

Switchgrass

Poaceae

Shrubs

Paspalumdilatatum

Sticky heads

Poaceae

Shrubs

Pennisetumpurpureum

Elephant grass

Poaceae

Herbs

Phyllanthus amarus

Hurricane weed

Phyllanthaceae

Herbs

Phyllanthus emblica

Amla

Phyllanthaceae

Tree

Phyllanthus mirabilis

Bhuiamla

Phyllanthaceae

Shrubs

Phyllanthus niruri

Lokhi

Phyllanthaceae

Herbs

Pieris japonica

Lily of the valley bush

Ericaceae

Shrubs

Poatrivialis

Rough bluegrass

Poaceae

Shrubs

Pouzolziamixta

Soap nettle

Urticaceae

Shrubs

Schima wallichii

Needlewood

Theaceae

Tree

Senecio vulgaris

Groundsel

Asteraceae

Herbs

Setariasphacelata

Bristlegrass

Poaceae

Herbs

Setariaviridis

Green bristlegrass

Poaceae

Shrubs

Smilax aspera

Rough bindweed

Smilacaceae

Shrubs

Smilax zeylanica

Malayalam

Smilacaceae

Herbs

Solanum dulcamara

Woody nightshade

Solanaceae

Shrubs

Tetraceravolubilis

Hojachigue

Dilleniaceae

Shrubs

Trichosanthestricuspidata

Bitter snake gourd

Cucurbitaceae

Shrubs

Verbena officinalis

Holy herb

Verbenaceae

Herbs

Figure 2: Life form analysis from the study

Figure 3: Dominant families from the study

Number of species

Figure 4. Plant species under different groups

Number

Of

Species

CONCLUSION

The result of the present study revealed the knowledge about the medicinal plant species; the preservation of this knowledge appears to be the result of continued reliance of local communities, medicinal and edible plants. The medicinal plants used by the local people to cure different diseases also reveal that many wild species are under growing pressures from various anthropogenic factors. The natural health care system is getting a great attention these days. Therefore, documentation of information on indigenous knowledge and practices will help in conserving the knowledge.

It is evident from the above study that the Ri-Bhoi district of Meghalaya is rich in plant wealth, which indicates the importance for biodiversity conservation. It has been observed that maximum plant species are important. The frequent flood, grazing, fishing etc. are some of the factors, which is responsible for the depletion of the biodiversity. But there is also need to give attention for conservation of the plant resources in the study site because without undisturbed characteristic flora of the area it is simply not possible for the native fauna of the study site to persist for longer period.

The study also showed that the area has plenty of medicinal plants to treat a wide spectrum of human ailments. Lack of compassion in the present generation has wiped out many rich wild flora of the area. It is an urgent need to take action and create awareness about the usefulness of the flora so that people can save this wealth. Cultivation of threatened medicinal plants should be encouraged by the local community in order to relieve pressure on these plants. It is hoped that this study and will provide a useful information on the conservation and sustainable use of the natural resources of the area.

REFERENCE

Ahmad, I., M.S.A., Ahmad, M., Ashraf, M., Hussain and M.Y. Ashraf., 2011. Seasonal variation in some medicinal and biochemical ingredients in a potential medicinal plants Mentha longifolia (L.) Huds. Pak. J. Bot., 43(SI): 69-77.

Ahmedullah, M.& Nayar, M.P., 1986. Endemic Plants of the Indian Region, Peninsular India , Volume I. Botanical Survey of India, Calcutta, India.

Arnachalam, K. and Parimelzhagan, T., 2011. Ethnomedicnal Onservation among Hooralis Tribes in Kadambur Hills, (Kalkadmbur) Erode District, Tamil Nadu. Global Journal of Pharmacology, 5, 117-121

Balakrishnan, N.P., 1983. Flora of Jowai and vicinity, Meghalaya. Botanical Survey of India, Howrah, Vol 2: 434-443

Beddome, R.H., 1868-1874. Icones Plantarum Indiae Orientalis .Ganz Brothers, Madras.

Bora, P.J., 1999. Flora and biodiversity of Pabitora Wildlife Sanctuary, Assam in North East India (Ph. D. Thesis N.E.H.U., Shillong).

Davis, G.& Richardson, D., 1995. Mediterranean Type Ecosystems: The Function of Biodiversity .Springer, Berlin, Germany.

Drude, D., 1980. Handbuch Der Pflanzengeographite. Stuttgart J. Engelhorn, p. 582.

Ellis, J.L., 1987. Flora of Nallamalais.Vol. 1-2, 7 Botanical Survey of India, Calcutta.

Gadgil, M.& Vartak,VD., 1975. Sacred groves of India: a plea for continued conservation. J Bombay Nat His Soc 72: 314-320.

Gadgil, M,, Utkarsh, G. & Pramod, P., 1996. Genus Ficus – Trees of Life. The Hindu Sur Env 171-175.

Gamble, J.S., 1915-36 Fischer CEC. Flora of the Presidency of Madras.Vol. 1-3, Adlard and Son Ltd., London.

Gamble, J.S.& Fischer, C.E.C.,1915-1935. Flora of the Presidency of Madras , Vols. 1-3. Adlord and Sons Ltd., London.

Ganesh, T., Ganeshan, R., Soubadra, Devy, M., Davidar, P.& Bawa, K.S.,1996.Assessment of plant biodiversity at a mid-elevation evergreen forest of Kalakad-Mundanthurai Tiger reserve, Western Ghats, India. Cur Sci 71: 379-392.

Gogoi, A. B., 1997. The study of the Herbaceous Plants of Tinsukia sub-div.Of Dibrugarh District, Assam with ref.To their Taxonomy & Utilization. M.Phil. Dissertation, Gauhati University.

Haridasan, K.& Rao, R.R , 1985-1987. Forest Flora of Meghalaya , 2 Volumes. Bishen Singh Mahendra Pal Singh, Dehra Dun, India.

Hazra, P.K., 1981. Nature and Conservation in Khasi folk beliefs and Taboos. Pp. 149-152. In S.K. Jain (ed.) Gilmpses of Indian Ethnobotany.Oxford and IBH Publishing Co. New Delhi.

Henry, A.N., Chithra, V.N. and Balakrishnan, P., 1989.Flora of Tamil Nadu India. Series 1: Analysis. Vol. III. Botanical Survey of India, Coimbatore.

Henry, A.N., Kumari, G.R. and Chitra, V., 1987.Flora of Tamil Nadu India. Series 1: Analysis. Vol. 2, Botanical Survey of India, Coimbatore.

Henry, AN, Chitra, V. & Balakrishnan, NP.,1989. Flora of Tamil Nadu, India , Series 1: Vol. 2. Botanical Survey of India. Southern Circle, Coimbatore.

Hooker, JD.,1872-1897. Flora of British India , Volumes 1-7. L. Reeve and Company, Ashford, Kent, U.K.

Humaira Shaheen, Rahmatullah Qureshi, Iram Zahra, Mubashrah Munir and Muhammad Ilyas., July., 2010, Floristic Diversity of Santh Saroola, Kotli Sattian, Rawalpindi, Pakistan

International Union for Conservation of Nature.,1980. World Conservation Strategy .IUCN-UNEP-WWF, Gland, Switzerland.

International Union for Conservation of Nature.,1994. R e d List Categories.IUCN Publications, Switzerland.

Israel, EDOI, Viji, C.& Narasimhan, D.,1997. Sacred groves: traditional ecological heritage. Int J Eco Env Sci 23: 463-470.

Jain, S.K. and Rao, R.R., 1976. A Hand Book of Field and Herbarium Methods. Today and Tomorrow's Printers and Publishers, New Delhi.

Jeeva, S & Anusuya, R.,2005.Ancient ecological heritage of Meghalaya. Magnolia 3: 20-22.

Jeeva, S, Kingston, C, Kiruba, S, Kannan, D, Jasmine, T, Sawian & Venugopal, N., 2006. Sacred forests – treasure trove of medicinal plants: a case study from Trivandrum district of southern Western Ghats In: National Seminar on Medicinal, Aromatic and Spices Plants: Perspective and Potential. Organized by Indira Gandhi Krishi Vishwavidyalaya, TCB College of Agriculture and Research Station, Bilaspur, Chattisgarh.

Jeeva, S, Kiruba, S, Mishra, BP, Venugopal, N, Kharlukhi, L, Regini, GS, Das, SSM & Laloo,RC., 2005. Diversity of medicinally important plant species under coconut plantation in the coastal region of Cape Flora Fauna.

Jeeva, S, Mishra, BP, Venugopal, N.& Laloo, RC., 2005. Sacred forests: traditional ecological heritage in Meghalaya. J Scott Res Forum 1: 94-97.

Joseph, J., 1982. Flora of Nongpoh and vicinity. Government of Meghalaya, Shillong: 248- 254

Kabeer, A.K. and Nair, V.J., 2009.Flora of Tamil Nadu Grasses. Botanical Survey of India, Kolkatta

Kanjilal, U.N., Dey, H.N. and Das, P.C., 1940. Flora of Assam.Govt. of Assam, Shillong. Vol.4: 232-274

Kharkongor,P. & Joseph, J., 1981. Folklore medicobotany of rural Khasi and Jaintia tribes in Meghalaya. pp 124-136. In S.K. Jain (ed.) Glimpses of Indian Ethnobotany.Oxford and IBH Publishing Co. New Delhi.

Kumar and Bussmann., 2008, Journal of ethnobiology and ethnomedicine.

Kumar., 2009, An ethnobotanical study of medicinal plants used by the locals in Kishtwar, Jammu and Kashmir, India. Ethnobotanical leaflets

Long & Li., 2004, studied the folklore of 66 medicinal plant species traditionally collected and used by the Red headed Yao people

Mathew, K.M., 1983. The flora of Tamil Nadu Carnatic, Parts (1-3). Rapient Herbarium, Tirudhirapalli.

Mishra, B.P, Tripathi, R.S, Tripathi, O.P. & Pandey, H.N.,2003.Effect of disturbance on the regeneration of four dominant and economically important woody species in a broad-leaved subtropical humid forest of Meghalaya, northeast India. Cur Sci 84: 1449-1453.

Mohanan, M.& Henry, A.N.,1994. Flora of Thiruvananthapuram, Kerala . Bulletin of Botanical Survey of India , Calcutta, India, p. 621.

Myers, N.,1993. Questions of mass extinction. Biodiver Conser 2: 2-17.

Nair, K.K.N.& Nayar, M.P.,1986-1987. Floraof Courtallum , 2 Vol. Botanical Survey of India, Calcutta.

Nath,S.M., 1999. Floristic composition of Orang Wildlife Sanctuary of Assam: A Comprehensive study. (Ph.D. thesis, Gauhati University)

Nayar, M.P.,1996. Hot Spots of Endemic Plants of India, Nepal and Bhutan . Tropical Botanical Garden and Research Institute (TBGRI), Thiruvananathapuram, Kerala, p. 252.

Nayar, M.P.,1959. The vegetation of Kanyakumari district. Bull Bot Surv India 1: 122-126.

Padalia, H., Chauhan, N., Porwal, M.C. and Roy, P.S., 2004.Phytosociological Observations on Tree Species Diversity of Andaman Islands, India. Current Science, 87, 799-806

Paria, N., 2005. Medicinal Plant Resources of SouthWest Bengal.Kolkata: Directorate of Forests, Govt. of West Bengal.

Parthasarathy, N., 1999 Tree Diversity and Distribution in Distributed and Human-Impacted Sites of Tropical Wet Evergreen Forest in Southern Western Ghats, India. Biodiversity and Conservation, 8, 1365-1381

Phillips, O.L., Martinez, R.V., Vargas, P.N., Monteagudo, A.L., Zans, M.E.C., Sanchez, W.G., Cruz, A.P., Timana, M., Yli-Halla, M. and Rose, S., 2003.Efficient Plot-Based Floristic Assessment of Tropical Forests. Journal of Tropical Ecology, 19, 629-645

Pitchairamu, C, Muthuchelian, K and Siva, N., Floristic inventory and quantitative vegetation analysis of tropical dry deciduous forest in Piranmalai forest, Estern Ghat, Tamilnadu, India, Ethnobot. leaflets. 12(1), 204 - 216, (2008).

Poongodi, A., Thilagavathi, S., Aravidhan, V. and Rajendran, A., 2011. Observation on Some Ethnomedicinal Plants in Sathyamangalam Forests of Erode Distict, Tamil Nadu, India. Journal of Medicinal Plants Research, 5, 4709-4714.

Prof. Dr. Sahadeo Kondaji Aher., 2010-2013, Study of Floristic Diversity of Parner Tehsil

Qureshi, R., 2004. Floristic and Ethnobotanical Study of Desert Nara Region, Sindh.Department of Botany, Shah Abdul Latif University, Khairpur, Sindh, Pakistan. Ph.D. Thesis, Vol. I: 1-300.

Qureshi, R.A., M.A. Ghufran, S.A. Gilani, Z. Yousaf, G. Abbas.and G. Batool., 2009. Indigenous medicinal plants used by local women in southern Himalayan regions of Pakistan Pak. J. Bot., 41: 19-25

Rajendran, A., Aravindhan, V. and Sarvalingam, A., 2014. Biodiversity of the Bharathiar University Campus, India: A

Floristic Approach. International Journal of Biodiversity and Conservation, 6, 308-319.

Rao, R.R.& Neogi, B., 1980. Observation on the Ethnobotany of Khasi and Garo tribes in Meghalaya (India) J Econ.Tax. Bot. Vol. 1: 157-162

Rao, R.R., 1981. Ethnobotanical studies of the flora of Meghalaya: Some interesting reports of herbal medicines. Pp 137-148. In S.K. Jain (ed.) Glimpses of Indian Ethnobotany. Oxford and IBH Publishing Co., New Delhi.

Revathi, P. and Parimelazhagan, T., 2010. Traditional Knowledge on Medicinal Plants Used by the Irula Tribes of Hasanur Hill, Erode District, Tamil Nadu. Ethnobotanical Leaflets, 14, 136-160

S.D. Vediya.and H.S. Kharadi., Sept 2011, International Journal of Pharmacy and Life Sciences. Floristic diversity of Isari zone, Megharj range forest District Sabarkantha, Gujarat, India

Santapau, H., 1955.Instructions for Field Collators of the Botanical Survey of India.Ministry of Natural Resources & Scientific Research, New Delhi.

Sarkar, S., 1993: Studies on Herbaceous Plants of Karbi-Anglong District of Assam with reference to their Taxonomy & Economic Utilisation. Ph. D.Thesis (Unpublished), Gauhati University.

Sharma, M.P, Ahmad, J., Flora of Mewat (Gurgaon District) Haryana. Journal of Economic and Taxonomic Botany 1995; 19(1):55-62. 24.

Sharma, M.P., Alternanthera paronychioides (Amaranthaceae) New Plant Record for Delhi. Journal of Economic and Taxonomic Botany 1994; 18(2):395. 23.

Sharma, M.P., Flora of Delhi: New Plant Record for Delhi, Journal of Economic and Taxonomic Botany. 1997; 21(1):245-246.

SilarMohammed, M., S.A. Rasheed and S. Maqbool Ahamed., 2009. Indian Journal of Applied and Pune.Biology, 24(1): 183-186

Sindhu, R., Rajendran, A. and Jayanthi, P., 2012.Herbaceous Life Forms of Maruthamalai Hills, Southern Western Ghats, India. International Journal of Medicinal and Aromatic Plants, 4, 625-631.

Sindhu, S., Uma, G. and Kumudha, P., 2012. Survey of Medicinal Plants in Chennimallai Hills, Erode Districts, Tamil Nadu. Asian Journal of Plant Science and Research, 2, 712-717.

Singh, R., 1993. Systematic Study on the Dicotyledonous Plants of Lakhimpur District (Undivided), Assam Ph.D. Thesis (Gauhati University).

Srivastava, S.K., 2011.Plant Diversity and Conservation Strategies of Uttar Pradesh. Phtotaxonomy, 11, 45-62.

Staden., 2008, Journal of ethnopharmacology

Thilagavathi, S., 2011. Flora of Xerophytes in Sathyamangalam Reserve Forest, Erode District, Tamil Nadu. M.Sc., Dissertation, Bharathiar University.

Vediya, SD. Kharadi,HS., Study of plant diversity in Megharj range forest District, Sabarkantha, North Gujarat, India, International Journal of Pharmacy & Life Sciences. 2011; 2(7):903-908.

Virupaksha, K., Shrihari, S., Madhyastha, M.N. and Babu Narayan, K.S., A new diversity indere for evaluation of environmental quality, J. Environ. Res. Develop., 4(1), 254257, (2009).

Viswanathan MV, Singh HB. Bhawat, PR., Additions to the Flora of Delhi. Journal of Economic and Taxonomic Botany. 1982; 14(3):579-580

List of Contributors

1. Soil Resources and Land use systems: Study in Ri-bhoi District, Meghalaya

L. S. Chanu, I. Laskar and S.I. Bhuyan

2. Plant Resources of Some Tropical Forest Stands of Ri-bhoi District, Meghalaya

I. Laskar and S.I. Bhuyan

3. A Comparative study of Soil Characteristics Found in Different Land use systems of Ziro Valley, Arunachal Pradesh

Hage Yama, I. Laskar and S.I. Bhuyan

www.ingramcontent.com/pod-product-compliance
Lightning Source LLC
La Vergne TN
LVHW050408160726
843469LV00041B/995